Guttmann / Scholz-Strasser (Eds.)
Freud and the Neurosciences

Freud and the Neurosciences
From Brain Research to the Unconscious

Edited by
Giselher Guttmann and Inge Scholz-Strasser

with contributions by
Oliver W. Sacks, Giselher Guttmann, Harald Leupold-Löwenthal,
Malcolm Pines, Cornelius Borck, Morris N. Eagle,
and Detlef B. Linke

VERLAG DER
ÖSTERREICHISCHEN AKADEMIE DER WISSENSCHAFTEN
VIENNA 1998

This publication has been supported by Society of Friends of the Sigmund Freud Museum, Vienna and Österreichische Akademie der Wissenschaften

Die Deutsche Bibliothek - CIP-Einheitsaufnahme
Freud and the neurosciences : from brain research to the unconscious / Giselher Guttmann ; Inge Scholz-Strasser (ed.). With contributions by Oliver Sacks ... - Wien : Verl. der Österr. Akad. der Wiss., 1998
ISBN 3-7001-2740-5

Translations from the German (except pp. 11–22, 47–55, 87–101) and editorial assistance:
Maria E. Clay
Project coordination: Katharina Murschetz assisted by Brigitte Flatschacher and Ralf Rother

All rights reserved. No part of this publication may be reproduced in any form or by any means graphic, electronic or mechanical; including photocopying, recording, taping or information storage and retrieval systems, without the written permission of the publisher.

Copyright for illustrations by S. Freud: A. W. Freud *et al.*/Mark Paterson & Associates

Cover design: Elke Salzer

Copyright © 1998 by
Österreichische Akademie der Wissenschaften/Austrian Academy of Sciences Press
P.O.Box 471, A-1011 Vienna
Tel. 00 43-1-515 81-405, Fax 00 43-1-515 81-400
e-mail: verlag@oeaw.ac.at

Printed by Grasl Druck & Neue Medien, A-2540 Bad Vöslau, Austria

Contents

Giselher Guttmann, Inge Scholz-Strasser: Preface 7

Oliver W. Sacks: Sigmund Freud: The Other Road 11

Giselher Guttmann: From the Sum of Excitation to the Cortical
DC Potential. Looking Back a Hundred Years 23

Harald Leupold-Löwenthal: Freud as a Neurologist 37

Malcolm Pines: Neurological Models and Psychoanalysis 47

Cornelius Borck: Visualizing Nerve Cells and Psychical Mechanisms.
The Rhetoric of Freud's Illustrations 57

Morris N. Eagle: Freud's Legacy. Defenses, Somatic Symptoms and
Neurophysiology 87

Detlef B. Linke: Discharge, Reflex, Free Energy and Encoding 103

List of Facsimiles 109

Index of Proper Names 111

Biographies 115

Preface

For I am actually not at all a man of science, not an observer, not an experimenter, not a thinker. I am by temperament nothing but a conquistador – an adventurer, if you want it translated – with all the curiosity, daring, and tenacity characteristic of a man of this sort.

Sigmund Freud to Wilhelm Fliess
February 1, 1900

Freud and the Neurosciences. From Brain Research to the Unconscious – that was the challenge posed by a symposium organized in spring 1997 by the Sigmund Freud Society in conjunction with the Austrian Academy of Sciences. The present volume presents revised versions of the papers from that conference, which was opened by the President of the Academy, Werner Welzig, and presided over by August Ruhs. The publication of these papers as a book has also been made possible by the Academy of Sciences in Vienna.

What was our motivation in taking a closer look at Sigmund Freud's "preanalytic period"? Is it a matter of pointing out the existence of two intellectually disparate researchers of whom only the second is known to most people? Or an attempt to provide evidence of a shift in Freud's life's work analogous to our view of Wittgenstein I and Wittgenstein II?

In fact our intention was a different one. Looking at Freud's early scientific work was intended to show the epistemological foundations that determined his entire oeuvre, even though at first glance one might not expect that they would also have provided a suitable basis for psychoanalysis. However, it was a decidedly physicalistic-scientific concept of the world with which the young Freud was first confronted by his teachers and which he readily accepted and, this is our thesis, never really abandoned or modified in any decisive way.

A topical incentive for this conference has been the current "Decade of the Brain" proclaimed in 1990 upon the initiative of the Library of Congress together with the National Institute of Mental Health and aimed at encouraging a public discussion of the topicality of brain research at the end of the 20th century. The point of departure was to be Freud's little-known early scientific research in neuroscience, first as a student and subsequently as a physician at the Vienna General Hospital. While still a student he published his first research papers not on the fringe of the scientific community but at the very center of academic research, in the *Sitzungsberichte der Akademie der Wissenschaften Wien*. These papers reflect Freud's early scientific career which led him from his physiological research on eels to the river crayfish and its nervous system. As a consequence of the neurological studies which he had begun in the laboratory of his teacher, the Academy member Ernst von Brücke, Freud developed his earliest psychological theory. Although he soon rejected this scientific model which attempted to explain the psychical apparatus by means of the physiology of the brain, his scientific interest was henceforth determined by his search for the precise structure of the psyche.

The authors of the present book deal with Freud's early research papers with a two-fold purpose: to place them in the tradition of the formation of the theories of the Second Vienna Medical School and to pose the question of the organizational structures of the brain and hence the basis of neuronal and psychological activity that has been of undiminished interest ever since the controversial fundamental neurological research carried out around the turn of the century. Coming from different scientific disciplines, the authors nevertheless agree on the radicality of Freud's approach. Although his effort to find a closed physical-scientific explanatory model for all psychological phenomena in terms of a universalist scientific claim in his *Project for a Scientific Psychology* failed from the beginning, Freud's approach anticipated results that could be verified experimentally only by the possibilities of 20th century technology and that have led to new interpretative approaches in a number of disciplines, e.g. regarding the visualization of nerve cells and psychical mechanisms in the illustrations of Freud's works (Borck) and the acceptance of Freud's neurological models in psychiatry by his disciples Goldstein and Schilder (Pines). The papers by Sacks and Leupold-Löwenthal focus on the scientific importance of Freud's early case histories and studies, as reported e.g. in his articles on *Acute multiple neuritis of the spinal and cerebral nerves*, *The neuro-psychoses of defence (an attempt at a psychological theory of acquired hysteria, of many phobias and obsessions and of certain hallucinatory psychoses)* and *On Aphasia* and trace the development of Freud's concern with psychological phenomena. It was in his studies of hysteria that the change in paradigm occurred which marks the transition from neurophysiology to psychoanalysis. Concepts such as regression, borrowed from Hughlings Jackson's theoretical model of the functioning of the brain, were used by Freud to explain important psychological phenomena in his psychoanalytic model of the instincts, just as other concepts from brain research also found their way into the interpretational models of psychoanalysis. Hence there is justification for the relationship which Eagle establishes between the Freudian concept of "repression" and the current concept of a "repressive style" personality with its highly differentiated effects upon somatic functions which he discusses on the basis of empirical data.

Linke's paper, on the other hand, addresses a decisive point of interface between the neurosciences and psychoanalysis by reflecting upon the energetic approach so important to Freud, from the viewpoint of today's neuroscience which is dominated by interpretational efforts based exclusively on the aspect of information processing – probably influenced by the computer metaphor of brain function. Guttmann establishes a bridge to state-of-the-art brain research by tracing the development from Freud's concept of the sum of excitation within his physicalistic world view to the recording of the cortical DC potential made possible by today's technology.

When Freud's scientific approach led him away from neurohistology and clinical neurology he entered new territory for which he had to create a new language, since no adequate concepts were available. This language was largely metaphorical and could therefore also be easily applied to quite different areas of cultural science. Nevertheless the concepts of psychotherapy are based on exactly the same scientific foundations that had determined Freud's early neuroscientific research, and the same epistemological approach seems to have determined also all his later work.

Not everybody may agree with this proposition. What is uncontested, however, is the fact that this epistemological orientation allows us to build a bridge from Freud's early neuroscientific work to today's research.

GISELHER GUTTMANN, INGE SCHOLZ-STRASSER

Freud: Über das Syrskische Organ etc.

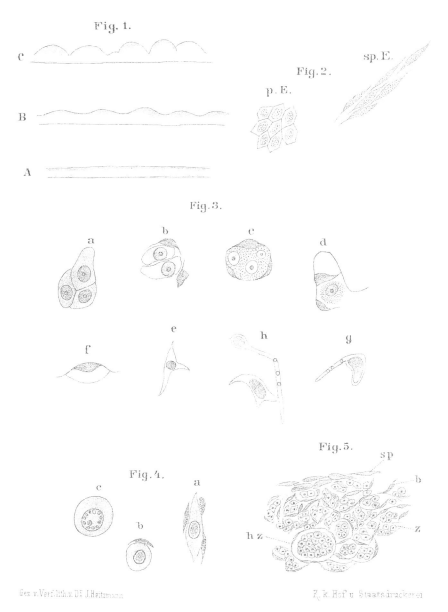

Erklärung der Abbildungen.

Fig. 1. Die hauptsächlichsten Formen des Lappenorgans. Schematische Zeichnung.
    A Lappenloses Organ.
    B Organ mit schmalen hyalinen Läppchen.
    C Entwickeltes Lappenorgan.

Fig. 2. Epitelien des Lappenorgans isolirt aus Müller'scher Flüssigkeit.
    p. E. polygonales Epitel.
    sp. E. Spindelzellen-Epitel.

Fig. 3. Inhaltszellen und Bindegewebskörper des Lappenorgans isolirt, aus Müller'scher Flüssigkeit. Vergrössert gezeichnet nach Hartn. $4/8$,
    a drei Inhaltszellen
    b zwei Zellen von Bindegewebskörpern umringt.
    c Kerne in feinkörnigem Protoplasma von Bindegewebskörpern eingeschlossen.
    d zwei Bindegewebskörper mit einander verbunden, deren leistenförmige Fortsätze eine Zelle einrahmen.
    e Bindegewebszelle mit grossem Protoplasmasaum.
    f Bindegewebszelle mit ringförmiger Leiste.
    g Bindegewebskörper mit leistenförmigem Fortsatz.
    h Ungewöhnliche Form der Verbindung zweier Bindegewebskörper durch ein geknicktes Leistchen.

Fig. 4. Ungewöhnliche Zellen aus einem kleinen Lappen. *a* und *b* isolirt aus Müller'scher Flüssigkeit, *c* isolirt aus Überosmiumsäure. Die Zellen von spindelförmigen Körpern umgeben.

Fig. 5. Ansicht eines Stückchen vom freien Rande des Lappenorgans zwischen zwei kleinen Läppchen.
    sp. Spindelzellen.
    b. Bindegewebskörper.
    z. Zellen des Lappenorgans.
    hz. Zellen des Lappenorgans in Häufchen angeordnet.

Oliver W. Sacks

Sigmund Freud: The Other Road

It is making severe demands on the unity of the personality to try and make me identify myself with the author of the paper on the spinal ganglia of the petromyzon. Nevertheless I must be he, and I think I was happier about that discovery than about others since.

Sigmund Freud to Karl Abraham
September 21, 1924

Everyone knows Freud as the father of psychoanalysis, but most people know little about the twenty years (1876–1896) when he was primarily a neurologist and anatomist; Freud himself rarely referred to them in later life. Yet his "other," neurological, life was the precursor to his psychoanalytic one, and perhaps an essential key to it. An early and enduring passion for Darwin, Freud tells us in his *Autobiography* (allied with Goethe's essay on Nature), made him decide to become a medical student; and already in his first year at university he was eagerly attending courses on "Biology and Darwinism" as well as lectures by the physiologist Ernst Wilhelm von Brücke. Two years later, eager to do some real, hands-on research, Freud asked Brücke if he could work in his laboratory. Although, as Freud was later to write, he already felt that the human brain and mind might be the ultimate subject of his explorations, he was intensely curious, after reading Darwin, about the early forms and origins of nervous systems, and wished to get a sense of their slow evolution first.

Brücke suggested that Freud look at the nervous system of a very primitive fish, *Petromyzon*, the lamprey – in particular at the curious "Reissner" cells clustered about the spinal cord; these cells had attracted attention since Brücke's own student days forty years earlier, but their nature and function had never been understood. Freud was able to detect the precursors of these cells in the singular larval form of the lamprey, and to show they were homologous with the posterior spinal ganglia cells of higher fish – a significant discovery. (This so-called *Ammocoetes* larva of *Petromyzon* is so different from the mature form that it was long considered to be a separate genus, *Ammocoetes*.) He then turned to looking at an invertebrate nervous system, that of the crayfish. And while it was believed at this time that the nerve elements of invertebrate nervous systems were radically different from those of vertebrate ones, Freud was able to show that they were, in fact, morphologically identical – and thus that it was not the cellular elements which were different in primitive or advanced animals, but their *organization*. Thus there emerged, even in Freud's earliest researches, a sense of a Darwinian evolution whereby, using the most conservative means (the same basic anatomic cellular elements) more and more complex nervous systems could be built.[1]

It was natural that in the early 1880s – he now had his medical degree – Freud should move on to clinical neurology; but it was equally crucial to him that he continue his anatomical work too, looking now at human nervous systems, and this he did in the laboratory of the neuroanatomist and psychiatrist Theodor Meynert.[2] For Meynert (as for Flechsig, and other neuroanatomists at the time) such a conjunction did not seem at all strange. There was assumed to be a simple, almost mechanical, relation of mind and brain, both in health and disease; thus Meynert's 1885 magnum opus, entitled *Psychiatry*, bore the subtitle, *A Clinical Treatise on Diseases of the Fore-Brain*.

Although phrenology itself had fallen into disrepute, the localizationist impulse had been given new life in 1861, when the French neurologist Broca was able to demonstrate that a highly specific loss of function – of expressive language, a so-called expressive aphasia – followed damage to a particular part of the brain (the third frontal convolution) on the left side. Other correlations were quick in coming, and by the mid-1880s something akin to the phrenological dream seemed to be approaching realization, with "centers" being described for expressive language, receptive language, color perception, writing, and many other specific capabilities. Meynert revelled in this localizationist atmosphere – indeed he himself, after showing that the auditory nerves projected to a specific area of the cerebral cortex (the *Klangfeld*, or sound field), postulated that damage to this was present in all cases of sensory aphasia.

Freud, it was evident, was disquieted by this theory of localization, and at a deeper level, profoundly dissatisfied, too, for he was coming to feel that all

---

[1] It was generally felt at this time that the nervous system was a synctium, a continuous mass of nerve tissue, and it was not until the late 1880s and 1890s, through the efforts of Cajal and Waldeyer that the existence of discrete nerve cells – neurons – was appreciated. Freud, however, came very close to discovering this himself in his early studies. That he did not make his thoughts fully explicit at the time, and gain any of the fame that went to Waldeyer and Cajal, was a later source of mortification to him.

[2] Freud published a number of neuroanatomical studies while in Meynert's lab, especially focussing on the tracts and connections of the brainstem. He often called these anatomical studies his "real" scientific work, and he subsequently considered writing a general text on cerebral anatomy – but the book was never finished, and only a very condensed version of it was ever published, in Villaret's *Handbuch*.

localizationism had a mechanical quality, treating the brain, the nervous system, as a sort of ingenious but idiotic machine, with a one-to-one matching of elementary components and functions, denying it organization – and evolution and history.

During this period (1882–1885), he spent time on the wards of the Vienna General Hospital, and it was here that he honed his skills as a clinical observer and neurologist. His skill and narrative powers, his sense of the importance of a detailed case history, are evident in the clinicopathological papers he wrote at the time – a boy who died from a cerebral hemorrhage associated with scurvy; an eighteen-year-old baker's apprentice with acute multiple neuritis; and a thirty-six-year-old man with a rare spinal condition, syringomyelia, who had lost the sense of pain and temperature, but not the sense of touch (a dissociation caused by the very circumscribed destruction within the spinal cord).

In 1886, after spending four months with the great neurologist Charcot in Paris, Freud returned to Vienna to set up his own neurological practice. It is not entirely easy to reconstruct – from Freud's letters, or the vast numbers of studies and biographies about him – exactly what "neurological life" consisted of for him. He saw patients in his consulting room at 19 Berggasse, presumably a mixed bag of patients as might come to any neurologist then, or now: some with everyday neurological disorders – strokes, tremors, neuropathies, seizures, migraines; and others with functional disorders – hysterias, obsessive-compulsive conditions, and neuroses of various sorts.

He also worked at the Institute for Children's Diseases, where he held a neurological clinic several times a week. (His clinical experience there led to the books for which he became best known to his contemporaries, his three monographs on the infantile cerebral paralyses of children. These were greatly respected among the neurologists of his time, and are still, on occasion, referred to even now.)

As he continued with his neurological practice, Freud's curiosity, his imagination, his theorizing powers, were on the rise, demanding more complex intellectual tasks and challenges. His earlier neurological investigations, during his years at the General Hospital, had been of a very skillful but conventional type, but now as he pondered the much more complex question of the aphasias, he became convinced that one needed a different view of the brain. A more dynamic vision of the brain was taking hold of him.

It would be of great interest to know exactly how and when Freud discovered the work of the English neurologist Hughlings Jackson, who, very quietly, stubbornly, persistently, was developing an evolutionary view of the nervous system, unmoved by the localizationist frenzy all around him. Jackson, twenty years Freud's senior, had been moved to an evolutionary view of nature with the publication of Darwin's *Origin of Species* and Herbert Spencer's evolutionary philosophy, and in the early 1860s proposed a hierarchic view of the nervous system, picturing how it might have evolved from the most primitive reflex levels, up through a series of higher and higher levels, to those of consciousness and voluntary action. In disease, Jackson conceived, this sequence was reversed, so that a dis-evolution or dissolution or regression occurred, and with this a "release" of primitive functions normally held in check by higher ones.

While Jackson's views had first arisen in reference to certain epileptic seizures (we still speak of these as "Jacksonian" seizures today), they had then been applied to a variety of neurological diseases, as well as to an attempt to understand dreams, deliria, and insanities; and in 1879, Jackson applied them to the problem of aphasia, which had long fascinated those neurologists interested in higher cognitive function.

In his own 1891 monograph *On Aphasia*, a dozen years later, Freud repeatedly acknowledges his debt to Jackson. He considers in great detail many of the special phenomena which may be seen in aphasias – the loss of new languages while the mother tongue is preserved; the preservation of the most commonly used words, the most commonly practiced associations; the preservation of series of words (days of the week, etc.) more than single ones; the verbal substitutions or paraphasias which may occur; and above all the stereotyped, seemingly meaningless phrases which are sometimes the sole residue of speech, and which may be, as Jackson had remarked, the last utterance of the patient before his stroke. This for Freud, as for Jackson represented the traumatic "fixation" (and thereafter the helpless repetition) of a proposition or an idea – a notion which was to assume a crucial importance in his theory of the neuroses.

Moreover, Freud observed, many symptoms of aphasia seem to share associations of a psychological sort more than a physiological sort. Thus verbal errors in aphasias, paraphasias, arise from the associations of words – words of similar sound or similar meanings tending to be substituted for the correct word. Sometimes the substitution is of a more complex nature, and is not comprehensible as a homonym or homophone, but arises from some particular association which has been forged in the individual's past. (Here there is an intimation or premonition of Freud's later views of paraphasias and parapraxes as *interpretable*, as historically and personally meaningful.) Thus here, he emphasized, we must look not so much to the anatomy or physiology of the brain as to the nature of words and their associations (formal or personal), to the universes of language and psychology, the universe of *meaning*, if we wish to understand paraphasias.

His study of aphasia convinced Freud that the complex manifestations of aphasia were incompatible with any simplistic notion of word images lodged in the cells of a "center": "... *under the influence of Meynert's teachings the theory has been evolved that the speech apparatus consists of distinct cortical centres; their cells are supposed to contain the word images (word concepts or word impressions); these centres are said to be separated by functionless cortical territory, and linked to each other by the association tracts. One may first of all raise the question as to whether such an assumption is at all correct, and even permissible. I do not believe it to be."*

Instead of centers – static depots of images – Freud writes, one must think of "cortical fields," large areas of cortex endowed with a variety of functions, some facilitating, some inhibiting, each other. One cannot make sense of the phenomena of aphasia, he continues, unless one thinks of them in such dynamic, Jacksonian terms. Furthermore, such systems are not all at the same "level." There is, he proposes, a vertical structure to speech, with repeated representations or embodiments of function at many hierarchic levels – thus the "regressions" characteristic of aphasia, the (sometimes explosive) emergence of primitive, emotional speech, when higher-level, propositional speech has become impossible. Freud was the first to bring this Jacksonian notion of regression into neurology, and the first to import it into psychiatry; one feels, indeed, that Freud's use of the concept of regression in *Aphasia* paved the way to his much more extensive and powerful use of it in psychiatry. (One wonders what Hughlings Jackson might have thought of this vast and surprising expansion of his idea, but, though he lived to 1911, we do not know whether he had ever heard of Freud.)[3]

Freud indeed goes beyond Jackson when he implies that there are no autonomous, isolable centers or functions in the brain, but, rather, systems for achieving cognitive goals – systems which have many components, and which can be created or greatly modified by the experiences of the individual. Thus, given that literacy was not innate, it was not useful, he felt, to think of a "center" for writing (as his friend and former colleague Exner had postulated); one had, rather, to think of a system, or systems, being constructed in the brain as the person learns (this was a striking anticipation of the notion of "functional systems," developed by A. R. Luria, the founder of neuropsychology, fifty years later).

In addition to these empirical and evolutionary considerations, Freud lays great emphasis on epistemological considerations – the confusion of categories, as he sees

---

[3] If a strange silence or blindness attended Hughlings Jackson's work (his *Selected Writings* were only published in book form in 1931–32), a similar neglect attended Freud's book on aphasia. More or less ignored on publication, *On Aphasia* remained virtually unknown and unavailable for many years – even Head's great monograph on aphasia, published in 1926, makes no reference to it – and was only translated into English in 1953. Freud himself spoke of *On Aphasia* as "a respectable flop" and contrasted this with the reception of his more conventional book on the cerebral paralyses of infancy: *"There is something comic about the incongruity between one's own and other people's estimation of one's work. Look at my book on the diplegias, which I knocked together almost casually, with a minimum of interest and effort. It has been a huge success ... But for the really good things, like the 'Aphasia,' the 'Obsessional Ideas,' which threatens to appear shortly, and the coming aetiology and theory of the neuroses, I can expect no more than a respectable flop."*

it, the promiscuous mixing of the physical and mental: *"The relationship between the chain of physiological events in the nervous system and the mental processes is probably not one of cause and effect. The former do not cease when the latter set in ... but, from a certain moment, a mental phenomenon corresponds to each part of the chain, or to several parts. The psychic is, therefore, a process parallel to the physiological, 'a dependent concomitant'."*

Freud here endorses and elaborates Jackson's views: *"I do not trouble myself about the mode of connection between mind and matter,"* Jackson had written. *"It is enough to assume a parallelism."*

Psychological processes have their own laws, principles, autonomies, coherences; and these must be examined independently, irrespective of whatever physiological processes may be going on in parallel. Jackson's epistemology of parallelism or concomitance thus gave an enormous freedom to Freud to pay attention to the phenomena in unprecedented detail, to theorize, to seek a purely psychological understanding without any premature need to correlate them with physiological processes (though he never doubted that such concomitant processes must exist).

As Freud's views evolved in relation to aphasia, moving from a "lesion" or "center" way of thinking towards a dynamic view of the brain, there was a parallel movement in his views on hysteria. Charcot was convinced (and had first convinced Freud) that although no anatomical lesions could be demonstrated in patients with *hysterical* paralyses, there must be nonetheless a "physiological lesion" (an *état dynamique*) located in the same part of the brain where, in an established *neurological* paralysis, an anatomical lesion (an *état statique*) would be found. Thus, Charcot conceived, hysterical paralyses were identical, physiologically, with organic ones – and hysteria was to be seen, essentially, as a neurological problem, a special reactivity peculiar to certain pathologically sensitive individuals or "neuropaths."

For Freud, still steeped in anatomical and neurological thinking, and very much under Charcot's spell, this seemed entirely acceptable. It was extremely difficult for him to de-neurologize his thinking, even in this new realm where so much was mysterious. But within a year, he had become less certain. The whole neurological profession was in conflict over the question of whether hypnosis was physical or mental. In 1889 Freud paid a visit to Charcot's contemporary, Bernheim, in Nancy – Bernheim had proposed a psychological origin for hypnosis, and believed that its results could be explained in terms of ideas or suggestion alone – and this seems to have influenced Freud deeply. He had begun to move away from Charcot's notion of a circumscribed (if physiological) lesion in hysterical paralysis to a vaguer, but more complex sense of physiological changes distributed among several different parts of the nervous system, a vision which paralleled the emerging insights of *Aphasia*.

Charcot had suggested to Freud that he try to clarify the controversy by making a comparative examination of organic and hysterical paralyses.[4] This Freud was well-equipped to do, for when he returned to Vienna and set up his private practice, he started to see a number of patients with hysterical paralyses and of course, many patients with organic paralyses too, and to attempt to elucidate their mechanisms for himself.

By 1893, he had made a complete break with all organic explanations of hysteria: The lesion in hysterical paralyses must be completely independent of the nervous system, since in its paralyses and other manifestations hysteria behaves as though anatomy did not exist or as though it had no knowledge of it.

This, then, was the moment of crossover, of transit, when (in a sense) Freud would give up neurology, and notions of a neurological or physiological basis for psychiatric states, and turn to looking at these exclusively in their own terms. He was to make one final, highly theoretical attempt to delineate the neural basis of mental states – in his *Project for a Scientific Psychology* – and he never gave up the notion that there must ultimately be a biological "bedrock" to all psychological conditions and theories. But for practical purposes he felt he could, and must, put these aside for a time.

Though Freud turned increasingly to his psychiatric work in the late 1880s and 1890s, he continued to write occasional shorter papers on his neurological work. In 1888 he published the first description of hemianopsia in children; in 1895 a paper on an unusual compression neuropathy (meralgia paresthetica), a condition he himself suffered from, and which he had observed in several patients under his care. Freud also suffered from classical migraine and saw many patients with this in his neurological practice, and at one point apparently considered writing a short book on this subject too but, in the event, did no more than make a summary of ten "Established Points" which he sent to his friend Wilhelm Fliess in April 1895. There is a strongly physiological, quantitative tone in this summary, *"an economics of nerve-force"* which hinted at the extraordinary outburst of thought and writing which was to occur later in the year.

---

[4] The same problem was also suggested to Babinski, another young neurologist attending Charcot's clinics (and later to become one of the most famous neurologists in France). While Babinski agreed with Freud on the distinction between organic paralyses and hysterical ones, he later came to consider, when examining injured soldiers in World War I, that there was "a third realm": paralyses, anesthesias, and other neurological problems based neither on localized anatomical lesions nor on "ideas," but on broad "fields" of synaptic inhibition in the spinal cord and elsewhere. Babinski spoke here of a "syndrome physiopathique." Such syndromes, which may follow gross physical trauma or surgical procedures, have puzzled neurologists since Weir Mitchell first described them in the Civil War, for they may incapacitate diffuse areas of the body which have neither specific innervation nor affective significance.

It is curious and intriguing that even with figures like Freud, who published so much, the most suggestive and prescient ideas may appear only in the course of their private letters and journals; and no period in Freud's life was more productive of such ideas than the "secret" years in the mid-1890s when the thoughts he was incubating in himself were shared with no one except Fliess. Late in 1895, Freud launched on an ambitious attempt to bring together all his psychological observations and insights, and ground them in a plausible physiology, and at this point his letters to Fliess are exuberant, almost ecstatic: *"One evening last week when I was hard at work ... the barriers were suddenly lifted, the veil drawn aside, and I had a clear vision from the details of the neuroses to the conditions that make consciousness possible. Everything seemed to connect up, the whole worked well together, and one had the impression that the Thing was now really a machine and would soon go by itself ... Naturally I don't know how to contain myself for pleasure."* (October 20, 1895).

But this vision in which everything seemed to connect up, this vision of a complete working model of the brain and mind which presented itself to Freud with an almost revelatory lucidity, is not at all easy to grasp now (and indeed Freud himself was to write, only a few months later, "*I no longer understand the state of mind in which I hatched out the 'Psychology'.*").[5]

There has been intensive discussion about this *Project for a Scientific Psychology*, as it is now named (Freud's working title had been "A Psychology for Neurologists"). The *Project* makes very difficult reading, partly because of the intrinsic difficulty and originality of many of its concepts; partly because Freud uses outmoded and sometimes idiosyncratic terms which we have to translate into more familiar ones; partly because it was written at furious speed in a sort of shorthand; and partly because it may never have been intended for anyone's eyes but his own.

And yet the *Project* does bring together, or attempt to bring together, the domains of memory, attention, consciousness, perception, wishes, dreams, sexuality, defense, repression, and primary and secondary thought processes (as he was to call them) into a single coherent picture of mind, and to ground all of them in a basic physiological framework, constituted by different systems of neurons, their interactions and modifiable "contact barriers," and free and bound states of neural excitation.

Though the language of the *Project* is inevitably that of the 1890s, a number of its notions retain (or have assumed) striking relevance to many current ideas in neuroscience — and this has caused it to be reexamined by Karl Pribram and Merton

---

[5] Freud never reclaimed his manuscript from Fliess, and it was presumed lost until the 1950s, when it was finally found and published – although what was found was only a fragment of the many drafts Freud wrote in late 1895.

Gill, among others. Pribram and Gill, indeed, call the *Project* "a Rosetta Stone" for those who wish to make connections between neurology and psychology, and many of its ideas can be examined experimentally now in a way which would have been impossible at the time they were formulated.

The nature of memory occupied Freud from first to last – aphasia was seen as a sort of forgetting (and he had observed in his notes that an early symptom in migraine was often the forgetting of proper names); he saw a pathology of memory as central in hysteria *("Hysterics suffer mainly from reminiscences")*; and in the *Project* he attempted to explicate the physiological basis of memory at many levels. One physiological prerequisite for memory, he postulated, was a system of "contact barriers" between certain neurons – his so-called psi system (this was a decade before Sherrington gave synapses their name). Freud's contact barriers were capable of selective facilitation or inhibition, thus allowing permanent neuronal changes which corresponded to the acquisition of new information and new memories – a theory of learning basically similar to that which Donald Hebb was to propose in the 1940s, and which is now supported by experimental findings.

At a higher level, Freud regarded memory and motive as inseparable. Recollection could have no force, no meaning, unless it was allied with motive – the two had always to be coupled together; and in the *Project*, as Pribram and Gill emphasize, *"Both memory and motive are psi processes based on selective facilitation ... memories [being] the retrospective aspect of these facilitations; motives the prospective aspects."*[6]

Thus remembering, for Freud, though it required such local neuronal traces (of the sort we now call long-term potentiation), went far beyond them, and was essentially dynamic, transforming, reorganizing, throughout the course of life. Nothing was more central for the formation of identity than the power of memory; nothing more guaranteed one's continuity as an individual. But memories shift, and no one was more sensitive than Freud to the reconstructive potential of memory, the fact that memories are continually worked over and that their essence, indeed, *is* recategorization.

Arnold Modell has taken up this point both with regard to the therapeutic potential of psychoanalysis, and more generally, the formation of a private self. He quotes a letter which Freud wrote to Fliess in December 1896, in which he uses the term *Nachträglichkeit*, a term usually mistranslated, which Modell feels is most accurately rendered as "retranscription."

---

[6] The inseparability of memory and motive, Freud pointed out, opened the possibility of understanding certain *illusions* of memory based on intentionality: the illusion that one has written to a person, for instance, when one has not, but intended to; or that one has run the bath, when one has merely intended to do so. We never have such illusions unless there has been a preceding intention.

"As you know [Freud wrote] *I am working on the assumption that our psychic mechanism has come into being by a process of stratification, the material present in the form of memory traces being subjected from time to time to a* rearrangement *in accordance with fresh circumstances* – a retranscription. *Thus ... memory is present not once but several times over ... the successive registrations representing the psychic achievement of successive epochs of life ... I explain the peculiarities of the psychoneuroses by supposing that this translation has not taken place in the case of some of the material."*

The potential for therapy, for change, therefore, lies in the capacity to exhume such "fixated" material into the present so that it can be subjected to the creative process of retranscription and thus allow the stalled individual to grow once again and change.

Such remodellings are not only crucial, Modell feels, in the therapeutic process, but are a constant part of human life not only for day to day "updating" (an updating which those with amnesia cannot do) but for the major (and sometimes cataclysmic) transformations, the revaluations of all values (as Nietzsche would say) which are necessary for the evolution of a unique private self.

That memory does construct and reconstruct, endlessly, was a central conclusion of the experimental studies carried out by Frederic Bartlett in the 1930s. Bartlett shows in these, very clearly (and sometimes very entertainingly) how with retelling – either to others, or to oneself – a story, the memory of a story, or of a picture, gets continually changed. There is never, Bartlett feels, a simple mechanical reproduction of memory, but always an individual and imaginative reconstruction. Thus, he writes: *"Remembering is not the re-excitation of innumerable fixed, lifeless and fragmentary traces. It is an imaginative reconstruction, or construction, built out of the relation of our attitude towards a whole active mass of organized past reactions or experience, and to a little outstanding detail which commonly appears in image or in language form. It is thus hardly ever really exact, even in the most rudimentary cases of rote recapitulation, and it is not at all important that it should be so."*

In the last third of the twentieth century, the whole tenor of neurology and neuroscience has itself been moving to such a dynamic and constructional view of the brain, a sense that even at the most elementary levels – as, for example, in the "filling in" of a blind spot or scotoma, or the seeing of a visual illusion, as both Richard Gregory and V. S. Ramachandran have demonstrated – the brain constructs a plausible hypothesis or pattern or scene. Gerald Edelman, above all – drawing on the data of current neuroanatomy and neurophysiology, of embryology and evolutionary biology, of clinical and experimental work, and of synthetic neural modelling – has been creating a most detailed neurobiological model of the mind. And in this, the brain's central role is precisely one of constructing categories – first perceptual, then conceptual – and of an ascending process, a "bootstrapping," where through repeated recategorization at higher and higher levels, consciousness is finally achieved. Thus every perception is a creation, for Edelman, and every memory, all remembering is re-categorization, recreation.[7]

Such categories, for Edelman, depend on the "values" of the organism, those biases or dispositions (partly innate, partly learned) which, for Freud, were characterized as "drives," "instincts," and "affects." Thus "retranscription" becomes the model for the brain-mind's most fundamental activity. The attunement here between Freud's views and Edelman's is striking – and here, at least, one has the sense that psychoanalysis and neurobiology can be fully at home with one another, congruent and mutually supportive. And it may be that in this equation of "*Nachträglichkeit*" with "recategorization" we see a hint of how the two seemingly disparate universes – the universes of human meaning and of natural science – may come together.

Ernest Jones spoke of Freud as *"the Darwin of the mind,"* and Edelman, in his latest book on neural darwinism, dedicates it to the memory of Darwin and Freud. And this is not "just" the Freud of psychoanalysis, but the Freud who spent his first adult twenty years as a neuroanatomist, a clinical neurologist, and neuro-theorist – and laid the foundations upon which psychoanalysis could arise.

REFERENCES

EDELMAN, G. M. (1992). *Bright Air, Brilliant Fire: On the Matter of the Mind*, New York: Basic Books.
FREUD, S. (1925). *An Autobiographical Study*. In: *Standard Edition*, vol. XX, London: Hogarth Press, 1959, pp. 1–74.
FREUD, S. (1891). *On Aphasia*. Trans. by E. Stengel, London: Imago Publishing Co., 1953.
FREUD, S. (1895). *Project for a Scientific Psychology*. In: *Standard Edition*, vol. I, London, Hogarth Press, 1956, pp. 281–397.
JACKSON, J. H. (1996). *Selected Writings (1931–1932)*, ed. James Taylor, Gordon Holmes and F. M. R. Walshe. 2 vols., Nijmegen: Arts & Boeve.
JELLIFFE, S. E. (1937). "Sigmund Freud as a Neurologist," *Journal of Nervous and Mental Diseases*, 85, 696–697.
JONES, E. (1956). *Sigmund Freud: Life and Work*, vol. 1, *The Young Freud*. London: Hogarth Press.
LURIA, A. R. (1979). *The Making of Mind: A Personal Account of Soviet Psychology*. Eds. Michael and Sheila Cole, Cambridge: Harvard University Press.

---

[7] There are, of course, innumerable areas in neuroscience and neurobiology besides that of memory where Freud's influence, direct or indirect, has been profound. There are marked analogies between psychoanalysis and neuropsychology, as discussed by Solms and Saling. A. R. Luria himself was fascinated by Freud's work as a very young man, and wrote to him in 1922 regarding the new Psychoanalytic Society, which he had founded in Kazan. Luria was thrilled, he wrote in his autobiography, *The Making of Mind*, to receive a courteous reply from the great man, addressing him as "Mr. President," and giving him permission to translate some of his works into Russian.

Luria, A. R. (1922). *Psychoanalysis in Light of the Principal Tendencies in Contemporary Psychology,* Kazan (in Russian).

Modell, A. (1990) *Other Times, Other Realities: Towards a Theory of Psychoanalytic Treatment,* Cambridge: Harvard University Press.

Modell, A. (1993). *The Private Self,* Cambridge: Harvard University Press.

Pribram, K. H. and Gill, M. M. (1976). *Freud's "Project" Reassessed,* New York: Basic Books.

Sulloway, F. J. (1979). *Freud: Biologist of the Mind,* New York: Basic Books.

Solms, M. and Saling, M. (1986). "On Psychoanalysis and Neuroscience: Freud's Attitude to the Localizationist Tradition," *International Journal of Psycho-Analysis,* 67, 397–415.

Erklärung der Abbildungen.

Tafel I.

Fig. 1. Spinalganglion von Petromyzon, Goldpräparat, gezeichnet bei Hartnack Ocular 3, Objective 8 und X, Vergrösserung 520.

Spinalganglion mit 15 Zellen; 5 grössere und 1 kleine im ventralen, 8 mittelgrosse und 1 kleine im dorsalen Ast. Die Grössenunterschiede der dorsalen und ventralen Zellen sind hier nicht bedeutend. Von jeder der 13 Zellen erster und zweiter Grössenordnung sind beide Fortsätze zu verfolgen. Im dorsalen Ast eine Ranvier'sche Zelle RZ. Die letzten dorsalen Zellen etwas dislocirt.

Der centrale Fortsatz der Zelle n dislocirt und abgerissen.

Ein Zellkern ist nur in der Zelle c zu sehen, die übrigen Kerne durch die starke Färbung der Zellen unkenntlich.

Im ventralen Ast zwei breite durchziehende Fasern dz. Mittelstarke durchziehende Fasern reichlich in beiden Ästen. Angelehnte Fasern deutlich bei ang. Es sind zwei sympathische Äste vorhanden.

HW = hintere Wurzel.
$v\,A$ = ventraler Ast.
kz = Kleinzelle.
$d\,A$ = dorsaler Ast.
gz = Grosszelle.
zf = Zellenfasern.
ang, ang' = angelehnte Fasern.
$s\,A$ = sympathischer Ast.
dz = breite durchziehende Faser.
dz' = mittelstarke durchziehende Faser.
wf = um die Wurzel gewundene Faser.
RZ = Ranvier'sche Zelle.

Fig. 2. Spinalganglion von Ammocoetes, Goldpräparat, gezeichnet bei Hartnack 2/8, Obj. X konnte nicht angewendet werden, daher erschienen mehrere Zellen unipolar. Beim Zerdrücken des Spinalganglions zeigte es sich, dass alle Zellen, mit Ausnahme der Doppelzelle dpz, bipolar waren.

Die Doppelzelle zeigte, nachdem sie isolirt war, bei x einen zweiten centralen Fortsatz. Vergrösserung 305.

gf = Gefäss.
$s\,a$ = sympathischer Ast.
dz' = durchziehende Fasermasse.
hw = hintere Wurzel.
ang = angelehnte Faser.

Fig. 3. Spinalganglion, Goldpräparat, gezeichnet bei Hartnack 2/8, Obj. X konnte nicht angewendet werden. Beim Zerdrücken des Präparates konnte man die beiden Fortsätze der scheinbar apolaren Zelle *az* erkennen. Zwei Ranvier'sche Zellen *Rz* und *Rz'*, letztere mit sehr kurzem Fortsatz. Vergrösserung 435.

      hw = hintere Wurzel.
      gza = Grosszellenast.
      dz' = durchziehende Fasern.
      kza = Kleinzellenast.
      Rz, Rz' = Ranvier'sche Zellen.
      az = scheinbar fortsatzlose Zelle.
      ang = angelehnte Fasern, die im Bogen vom ventralen in den dorsalen Ast ziehen.

Fig. 4*A*. Ranvier'sche Zelle aus einem der letzten Spinalganglien, Goldpräparat.

      Fig. 4. *B—F*. Isolirte Zellen aus Spinalganglien, nach Bleistiftskizzen, die von den Präparaten gemacht worden waren, gezeichnet.

B bipolare Zelle mit Theilung des peripheren Fortsatzes.

C tripolare Zelle mit zwei centralen Fortsätzen aus einem Spinalganglion; eben solche Formen finden sich im Rückenmark.

D Ranvier'sche Zelle; der Fortsatz der Zelle *R* theilt sich bei *I*. Von den beiden Theilungsästen theilt sich der Ast *b* nochmals T-förmig bei *II*.

E Zwei anscheinend unipolare Zellen, deren Fortsätze sich vereinigen. (?)

F Ranvier'sche Zelle; der Fortsatz der Zelle *R* theilt sich zum ersten Male bei *I*, von den beiden Ästen theilt sich der eine *b* nochmals gabelig bei *II*.

GISELHER GUTTMANN

From the Sum of Excitation to the Cortical DC Potential Looking Back a Hundred Years

When Sigmund Freud began his medical studies in Vienna in 1873 he did not encounter an intellectual climate conducive to those ideas which we generally associate with the concept of "psychoanalysis". Would someone who was later to interpret the phenomenon of anxiety and offer explanations for both the conversion of affects with which the individual cannot cope into somatic symptoms and for the development of obsessional ideas not have to be at odds with the spirit then prevailing at the Vienna Medical Faculty? Even a fleeting glance at Freud's biography shows that this was not at all the case and that he was instead enthusiastic about a physiology oriented towards an ideal strongly influenced by natural science as Freud encountered it at Ernst von Brücke's Institute. Like Emil DuBois-Reymond, Hermann von Helmholtz and Carl Ludwig, Ernst von Brücke had been a disciple of Johannes Müller, who followed only hesitatingly the change of paradigm in the interpretation of nervous excitation processes as electric phenomena. His disciples had even founded a club [which later developed into the *Berliner Physikalische Gesellschaft*] in order to oppose Müller's vitalist concepts by offering scientific explanations. Müller's thesis that it would be impossible to determine the speed with which nervous excitation is transmitted was disproved by his disciple Helmholtz, who was the first to successfully measure the conduction speed of nerve impulses.

As taught in Vienna, physiology was epistemologically oriented towards the then leading science of physics and characterized by the consistent endeavor to offer closed physical and scientific explanation models for every phenomenon. DuBois-Reymond's statement [1918] can be considered representative of this attitude: "*Brücke and I pledged a solemn oath to put in power this truth: No other forces than the common physical chemical ones are active within the organism. In those cases which cannot at the time be explained by these forces one has either to find the specific way or form of their action by means of the physical mathematical method, or to assume new forces equal in dignity to the chemical physical forces inherent in matter, reducible to the force of attraction and repulsion.*"[1]

At Brücke's Institute Freud got to know Josef Breuer, a *Privatdozent*, who was a particularly consistent advocate of a physicalist physiology and whose ideas were to have a lasting influence upon Freud. Breuer had developed an energetic model in which the fluctuations of an energy at rest, an intracerebral tonic excitation, was supposed to be responsible for all psychological activity. As Breuer states in his contribution to the publication he wrote in cooperation with Freud [Freud & Breuer 1895]: *"We ought not to think of a cerebral path of conduction as resembling a telephone wire which is only excited electrically at the moment at which it has to function ... We ought to liken it to a telephone line ... Or better, let us imagine a widely-ramified electrical system for lighting and the transmission of motor power; what is expected of this system is that simple establishment of a contact shall be able to set any lamp or machine in operation. To make this possible ... there must be a certain tension present throughout the entire network of lines of conduction, and the dynamo engine must expend a given quantity of energy for this purpose. In just the same way there is a certain amount of excitation present in the conductive paths of the brain when it is at rest but awake and prepared to work."*

In view of this background it is not surprising that for a decade Freud's scientific work was devoted to neurohistology and clinical neurology. In his first published work *Über den Ursprung der hinteren Nervenwurzeln im Rückenmarke von Ammocoetes (Petromyzon planeri) [On the origin of the posterior nerve-roots in the spinal cord of Ammocoetes (Petromyzon planeri)]* [1877] Freud was able to show that in this primitive fish from the group of the lampreys [cyclostoma] the cell population discovered by Reissner, whose nature and function had never been understood, were spinal ganglia. In the course of his neurohistological work Freud even developed a new method of staining by means of gold chloride which he published in various journals and considered an important discovery. Even in the 1880s Freud still published such clinical and neurological works as *Ein Fall von Hirnblutung mit indirekten basalen Herdsymptomen bei Skorbut [A case of cerebral haemorrhage with indirect basal focal symptoms in a patient suffering from scurvy]* [1884] and *Akute multiple Neuritis der spinalen und Hirnnerven [Acute multiple neuritis of the spinal and cranial nerves* [1886].

Unexpectedly we come across Freud's 1894 publication entitled *Die Abwehrneuropsychosen. Versuch einer psychologischen Theorie der akquirierten Hysterie, vieler*

---

[1] Siegfried Bernfeld (1944) "Freud's Earliest Theories and the School of Helmholtz," *The Psychoanalytic Quarterly,* 13, 348. [German translation: Bernfeld, Siegfried and Cassirer-Bernfeld, Suzanne: *Bausteine der Freud-Biographik*. Ed., trans. and introd. by Ilse Grubrich-Simitis, Frankfurt/Main 1988, pp. 62 f]

Phobien und Zwangsvorstellungen und gewisser halluzinatorischer Psychosen [The neuropsychoses of defence (An attempt at a psychological theory of acquired hysteria, of many phobias and obsessions and of certain hallucinatory psychoses)][2] published in *Neurologisches Zentralblatt* Nos. 10 and 11 – which indicates an (apparently) sharp break in his point of view. In this study Freud reflects upon the fact that in some patients *"an occurrence of incompatibility took place in their ideational life"* when *"their ego was faced with an experience, an idea or a feeling which aroused such a distressing affect that the subject decided to forget about it because he had no confidence in his power to resolve the contradiction between that incompatible idea and his ego by means of thought-activity"*. If this forgetting is not successful, *"the affect which is attached to the idea ... must be put to another use."* It may manifest itself in somatic symptoms, a process for which Freud proposed the term "conversion". The *"affect, which has become free, attaches itself to other ideas which are not in themselves incompatible; and, thanks to this 'false connection', those ideas turn into obsessional ideas."* However, it is also possible that *"... the ego rejects the incompatible idea together with its affect and behaves as if the idea had never occurred to the ego at all. But from the moment at which this has been successfully done the subject is in a psychosis, which can only be classified as 'hallucinatory confusion'."*

These are bold interpretations which at first sight do not seem to fit the scientific and physicalist view of the world held by the Vienna School of Medicine to which Freud had so far felt committed and which do not seem to agree at all with the basic attitude of his earlier studies. Yet at the end of this publication Freud made a remarkable statement: *"I should like, finally, to dwell for a moment on the hypothesis which I have made use of in this exposition of the neuroses of defence. I refer to the concept that in mental functions something is to be distinguished – a quota of affect or sum of excitation – which possesses all the characteristics of a quantity (though we have no means of measuring it), which is capable of increase, diminution, displacement and discharge, and which is spread over the memory-traces of ideas somewhat as an electric charge is spread over the surface of a body.*

This hypothesis ... can be applied in the same sense as physicists apply the hypothesis of a flow of electric fluid. It is provisionally justified by its utility in co-ordinating and explaining a great variety of psychical states."

Thus Freud assumed an electric charge – the concept of "the sum of excitation" taken quite literally in the physical sense and explicitly declared "a hypothesis" which is only used to allow explanations and facilitate more accurate predictions. Should he therefore be considered a precursor of Constructivism? From a neuropsychological

---

[2] This topic was already mentioned a year earlier in a work Freud had written jointly with Breuer *Über den psychischen Mechanismus hysterischer Phänomene; Vorläufige Mitteilung [On the psychical mechanism of hysterical phenomena (Preliminary communication in collaboration with Dr. J. Breuer)]*, In: *Standard Edition*, vol. II, pp. 3–17.

viewpoint it is certainly a prophetic statement, for about a hundred years later we are in fact able to record cortical surface potentials that seem to be subject precisely to the laws postulated by Freud and to bear the character of such a "sum of excitation" as an expression of local excitability and response readiness. We owe this indicator of cortical activation, however, to another research motivation, namely the desire to be able to establish an "objective psychology" based on neurophysiology.

Gustav Theodor Fechner [1889] had already conceived of the possibility of gaining access to psychological phenomena through the biological foundations of experience. In the 1920s it seemed as though one could actually get closer to this bold idea of an "objective look into human experience", which must have seemed an entirely utopian concept in the 19th century. On 6 July 1924 the psychiatrist Hans Berger first observed that the human brain produces weak electric shifts in potential of a few millionths of a volt which, when recorded by suitable amplifiers, would yield a person's brain electric recording ("Hirnstrombild", electroencephalogram, EEG). Berger did not dare publish his findings until 1929, since he was uncertain as to whether these potentials that were barely measurable could really be regarded as a genuine biolectric activity [Berger 1929]. However, already a few years after the appearance of Berger's study, Hubert Rohracher in Vienna began to be interested in this phenomenon. Together with Ottenthal he built a functioning thermionic amplifier and already in the 1930s he was able to prove that the EEG reflects minute changes in our consciousness. [Rohracher 1935]

Whenever we concentrate on some task (he asked his subjects to solve e.g. mathematical problems) small waves occur in rapid succession. In a state of relaxation, on the other hand, slower waves predominate which Berger had already termed "alpha waves" because of their remarkable regularity. Whenever someone falls asleep this likewise manifests itself, as Rohracher was the first to show, in a marked shift in the EEG, namely the occurrence of even slower and larger waves that change in accordance with how deeply we sleep. In this way we obtain objective data about a person's state of consciousness, even being able to register objectively the current depth of sleep which naturally eludes our self-observation.

This seemed to represent a decisive step in the direction of a new "objective psychology", thus solving the problem of a psychology based solely on introspection. Already in 1937 Rohracher began his remarkable efforts to gain access to a person's experience by means of the electric phenomena of the brain [Rohracher 1937a and b], about which he said in a retrospective summary of his work on the EEG published since 1935 [1940]: *"... in this way the relationship between the brain electric phenomena and the psychological processes can also be clearly and precisely demonstrated."* As he stated in the same work, his central concern was the recording of the *"... electric phenomena of the brain underlying our conscious experience".*

Rohracher eventually even attempted to visualize brain electric concomitants of specific experiences and search for the effects of sensory stimuli upon the EEG. His pioneering work could, however, not be successful, since, as we now know in hindsight, a computer is required to record these potentials and this has only been available since the 1960s. It is only by averaging a sufficiently large number of responses to stimuli that a potential can be observed whose course obviously depends on the specific characteristics of the stimuli and shifts so regularly depending on their intensity that conclusions about the person's experience, i.e. "objective tests of sensory capability", are possible.

I myself was able to participate, together with Gidon Gestring, when the possibility was first utilized at Professor Burian's otolaryngology clinic in Vienna to test a person's hearing by means of these potentials and the first laboratory for EEG audiometry (computer audiometry) was established. Subsequently these perception potentials were used in numerous other applications and e.g. such hard to classify perceptions as the experience of pain could be tested objectively. Already at the time I was, however, concerned with the question of whether these potentials are simply an ultimate image of the stimulus occurring so that we, as it were, reconstruct in a very complicated way from brain electric phenomena what affects our sense organs from the environment, or whether instead these potentials correspond to our experience, i.e. may be regarded as correlates of psychic phenomena. In the late 1960s I finally succeeded in proving that cortical potentials triggered by acoustic stimuli in fact correspond to a person's momentary experience, even in those cases where a discrepancy is observed between stimulus and experience [Guttmann 1968]. Whenever a subject is presented with a longer series of objectively identical acoustic stimuli, occasionally some of them will be experienced as softer and others as louder. A selective analysis of such "discrepant situations" revealed that the potentials correspond to the experience: the brain electric responses to the stimuli perceived as louder correspond to those that are evoked by louder tones, while those experienced as softer resemble the potential of stimuli of lesser intensity!

After a "direct look at individual experience" – at least in the meaning of this formulation in terms of applications – had become possible in the 1960s by recording the correlates of experience and work had begun on the numerous practical applications, the Vienna Group focussed on another brain electric phenomenon which was to constitute the main emphasis of our research for the next twenty years: the cortical DC potential [or Slow Potential].

It had long been assumed that these potentials might be an indication of the excitability of a specific cortical area and that in contrast to the discrete impulse patterns by which nerve cells transmit information to specific circumscribed areas these shifts in potential extending over larger cortical regions represent an analog process which Pribram has appropriately termed "another language" of the brain [Pribram 1995].

This battery-like charge of the brain had first been described during the 19th century. In 1875 Caton reported about cortical DC potentials which he was able to observe by means of a string galvanometer. By the turn of the century Beck and Cybulski were able to observe that sensory stimuli triggered negative potential shifts in the respective cortical projective fields which are obviously indicators of local excitability. Nevertheless for a long time it was not possible to register extremely slow potential shifts with the thermionic amplifiers developed during the 1930s, so that these slow potential shifts were gradually forgotten.

They were only rediscovered in 1964 by Grey Walter with an experimental setup in which the brain electric correlate of an expectation could be observed. If an announcing stimulus [S1] is followed by a second stimulus [„imperative stimulus" – S2] at specific regular time intervals then a slow contingent negative shift, the Contingent Negative Variation [CNV] occurs, particularly if an active reaction to the second stimulus is required. This may be interpreted as the expression of a general attitude of expectancy and has therefore been termed Expectancy Wave [Walter 1964].

Its amplitude depends on the person's level of activation by the respective stimulus, e.g. the motivation with which he prepares for the reaction or how surprising the stimulus is to him. That can be proved e.g. by an experimental setup called an "Oddball CNV", using two different announcing stimuli – such as a high-pitched and a low-pitched tone – with one being presented more frequently than the other. In this case the announcing stimulus that is used less often [e.g. in 20 % of all cases] triggers a larger expectancy wave and the more frequent one a smaller one. In such a study carried out by us [Guttmann 1992] we were able to observe the remarkable fact that the sums of the amplitudes of the two corresponding potentials [i.e. the sums of the potential shifts for S1 with 20 % and S2 with 80 % likelihood of presentation] always result in the same total value, i.e. always reflect the same net amount of energy [Figure 1]. One might almost dare speak of a brain electric indicator of a "law of the preservation of mental energy", as Bernfeld had postulated in his "Libidometry" [Bernfeld & Feitelberg 1930].[3]

It has long been assumed that these local negative shifts are not only indicators of an increased activity of the respective cortical area but also influence a person's momentary capability and might somehow be especially connected with learning processes. Already in 1953 Rusinov was able to observe that a depolarization of the

---

[3] Quoted after: Bacher R. (1992): Libidometry. In: Fallend and Reichmayr (Eds.), *Siegfried Bernfeld oder die Grenzen der Psychoanalyse*

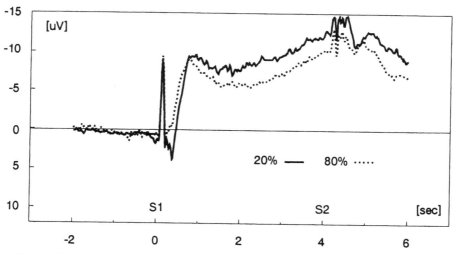

Figure 1: Frequent announcing signals trigger considerably smaller negativity than rare ones [here: An Oddball experiment with high-pitched (80 %) and low-pitched (20 %) tones].

cortex by an external source of electricity facilitates the formation of conditioned responses. The focus which he called "dominant focus of excitation" had put the respective area into a "state of learning readiness".

In many years of development work Herbert Bauer at our Institute successfully developed new methods of recording and amplifying such potentials, which permit the recording of cortical DC potential shifts by means of scalp electrodes. The patterns observed provide an image of the topical excitation level of the various cortical areas – is this an electrophysiological manifestation of the "tonic excitation"? – that can be visualized by means of suitable techniques and usually demonstrates an alternation between increased and decreased excitability.

The extent to which such fluctuations of the topical excitation level influence the learning performance could be demonstrated by us in an experimental setup which made it possible to present tasks by the computer precisely at the moment of a certain brain electric shift such as a spontaneously occurring negativity [Guttmann and Bauer 1984: "Brain Trigger Design"]. It turned out that whenever the test subject moved from a deactivated phase into a state of higher activation a marked improvement in learning capability occurred. The positive effect upon learning performance was so impressive that a practical application for the optimization of learning situations could be derived from this which has proved its effectiveness many times in a teaching model ["Learning under Self-Control" – LuS].

As an example of the precision with which the distribution of the DC potentials allows us to recognize the activity of the cortical areas responsible for a specific cognitive performance we present such an "excitation map" during the solving of spatial orientation tasks, from a recent study by Oliver Vitouch [1997] at our Institute. On a monitor the subjects were presented with views of two cubes, each with differently designed faces, the subjects' task being to decide whether the cube on the right could be a different view of the cube on the left – which was sometimes the case. To solve these tasks the subjects had to perform a "mental rotation" of the cube in question. It turned out that at the moment of solving the problem [in addition to the marked activation of the visual cortex which was the result of looking at the cubes that were being shown] a marked negative shift occurred in the right hemisphere precisely in the area that is responsible for the spatial orientation ability. In poor spatializers who had not done so well on an earlier spatial orientation test, i.e. for whom these tasks were more difficult, negative shifts were much more marked than in good spatializers. Thus the DC potential reflects in a way the extent to which the sum of excitation has increased in accordance with the degree of mental effort.

We were, however, faced with an entirely different excitation distribution whenever the same persons had to solve verbal analogies such as

walk : run = wind : storm

While the subjects were doing this verbal test (which was presented only to right-handed persons) a closely circumscribed left hemispheric negativity was seen exactly in the region of the speech center.

As a further illustration of the possible correspondence of DC potentials and Freud's "sum of excitation" let us mention two of our investigations of "altered states of

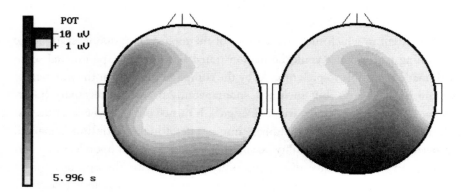

Figure 2: The distribution of DC potentials while solving verbal tasks (left) and spatial orientation tasks (right). In this as well as in the following figures dark zones show areas of increased and light zones areas of reduced cortical activity.

consciousness" which are particularly convincing examples of an application of this "access to psychological phenomena", since in these cases the subject himself can naturally not report on any topical experiences or the psychological changes involved.

For decades psychophysiological studies of hypnosis have been very impressive due to their remarkable discrepancies between the observations on the biological level and the subjective experience. Despite extremely dramatic changes in perception [such as insensitivity to pain which in some cases was even used as a means of analgesia during surgical interventions] the EEG showed hardly any interpretable pecularities. Together with Barolin and Gestring I myself carried out one of the first studies of acoustically evoked potentials under hypnosis and have had to observe – at the time to our disappointment but subsequently confirmed in numerous other studies – that during states of hypnotically induced hardness of hearing the amplitudes of the evoked potentials remain entirely unchanged despite marked changes in the threshold of perception.

Only recently have we been able to supply a tentative explanation for these paradoxes based on DC studies. In one of these studies we carried out a more subtle differentiation of the hypnotic state for the first time and put the subjects first into a state of a relaxed vacation situation. In the subsequent part of the experiment a situation was suggested to them in which motor activities predominated: the subjects were supposed to escape from a situation of pursuit – as though in a nightmare, their movements becoming ever slower and more difficult. After a while this negative idea was dissolved and the subjects were put back into a state of relaxed ease.

DC analysis provided precisely the expected picture. While the original relaxed state as well as the final phase showed a relatively uniform moderate activation of the entire cortex, during the hypnotic phase with the suggested movement only distinct local negative shifts were seen in precisely those areas in which one would expect an increased excitation in the imagined situations presented, namely in the occipital area responsible for optical activities and in the motor areas of the frontal brain [Figure 3]. Such a "spot activation" can obviously be triggered by hypnosis but seems to basically be the rule in cortical activation control by which every moment's information processing – and hence the selection of whatever we are conscious of at that moment – is controlled by a constant fading in and out of cortical areas.

For another study a state of consciousness was chosen in which an extreme change in the sum of excitation was to be expected: we were able to study a group of persons with many years of experience in Zen meditation who were willing to meditate in the laboratory. Zen meditation is a particularly interesting state of consciousness, since it is its express goal to induce a complete absence of thought while yet being in a state of concentrated wakefulness. Even under these unusual

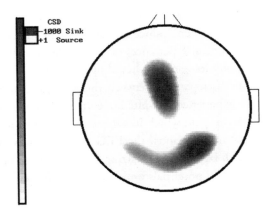

Figure 3: While under the suggestion of being pursued, two sharply delimited foci of excitation occur in the occipital and motor areas [current source density].

conditions it turned out that soon after the beginning of meditation a global and constant deactivation of the entire cortex occurred to an extent that can hardly ever be observed otherwise. This total reduction of electric charge corresponds precisely to the state of a total absence of thinking that is the goal of Zen – wakefulness without content, an extreme reduction of the sum of excitation.

DC recordings of a control group that had no experience with meditative techniques yielded a remarkable result. This group was played "relaxation music", as it is offered in many places today and seems to in fact trigger the experience of a state of contemplative peace. Remarkably, the DC potential distribution shows that there was no deactivation but instead a considerable cortical negative shift. Acoustic

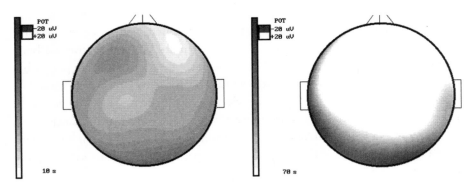

Figure 4: The distribution of the DC potential a few minutes after the beginning of a Zen meditation (left) and 70 seconds later (right). The dark areas once again show the activated regions and the light colored areas the deactivated regions.

stimulation induces activation and is in any case a far cry from the mental "emptiness" achieved by Zen meditation.

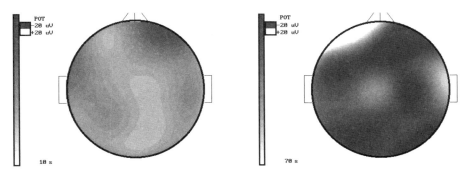

Figure 5: The analogous shifts while listening to relaxation music (at the beginning and 70 seconds later) which obviously did not effect any deactivation.

The electric component is no doubt only one aspect in which the local cortical activation is expressed. One decisive advantage of its observation is the fact that even very brief potential shifts can be observed that occur in fractions of seconds, whereas other imaging processes of undisputed importance such as Positrone Emission Tomography [PET] allow us to recognize the functional state of the brain only during a longer period of time. In addition, unlike other imaging processes DC registrations are not limited to the laboratory. As part of a field study we recently even installed the recording devices in a rally car and were thus able to study – with remarkable results – the brain electric shifts of drivers and passengers while they were driving along a high speed passage.

From a neuropsychological point of view we thus seem to find the "hypothesis" of a "sum of excitation", which was first postulated by Freud as *"capable of increase, diminution, displacement and discharge, and which is spread over the memory-traces of ideas somewhat as an electric charge is spread over the surface of a body"* quite wonderfully confirmed more than a hundred years later. Undoubtedly physical "hypotheses" of this kind largely continued to determine also Sigmund Freud's later thinking.

REFERENCES

BACHER, R. (1992). Libidometrie. In: Fallend, Karl and Reichmayr, Johannes (Eds.): *Siegfried Bernfeld oder die Grenzen der Psychoanalyse*, Basel: Stroemfeld/Nexus.

BAROLIN, G., GESTRING, G. and GUTTMANN, G. (1968). "The effects of hypnotic suggestions on the acoustic evoked potential," *EEG Clin. Neurophysiol.*, 24.

BAUER, H., REBERT, Ch., KORUNKA, Ch. and LEODOLTER, M. (1997). "Rare events and the CNV – the oddball CNV," *International Journal of Psychophysiology*, 13, 51–58.

BAUER, H. (1998). Slow potential topography. *Behavior Research Methods, Instruments, & Computers*, 30 (1), 20–33.

BERGER, H. (1929). "Über das Elektroencephalogramm des Menschen. 1. Mitteilung," *Archiv Psychiatrisches Nervenkrankenhaus*, 87, 527–570.

BERNFELD, S. (1944). "Freud's Earliest Theories and the School of Helmholtz," *The Psychoanalytic Quarterly*, 13, 348.

BERNFELD, S. and CASSIRER-BERNFELD, S. (1988). *Bausteine der Freud-Biographik*, Frankfurt am Main: Suhrkamp-Taschenbuch Wissenschaft.

FECHNER, G. T. (1889). *Elemente der Psychophysik*, Leipzig: Breitkopf & Härtel.

FREUD, S. (1877). "Über den Ursprung der hinteren Nervenwurzeln im Rückenmarke von Ammocoetes (Petromyzon Planeri)," *Sitzungsberichte der Akademie der Wissenschaften Wien* (Math.-naturwiss. Cl.), III. Abt., vol. 75, 15–27.

FREUD, S. (1884). "Ein Fall von Hirnblutung mit indirekten basalen Herdsymptomen bei Skorbut", *Wiener Medizinische Wochenschrift.*, vol. 34, No. 9, 244–246, and No. 10, 276–279.

FREUD, S. (1886) "Akute multiple Neuritis der spinalen und Hirnnerven," *Wiener Medizinische Wochenschrift*, vol. 36, 168–172.

FREUD, S. and BREUER, J. (1893). On the Psychical Mechanism of Hysterical Phenomena. Preliminary Communication. In: *Standard Edition*, vol II, London: Hogarth Press 1955, pp. 3–17.

FREUD, Sigmund and BREUER, J. (1895). Studies on Hysteria. In: *Standard Edition*, vol. II, London: Hogarth Press 1955.

FREUD, S. (1894) "Die Abwehrneuropsychosen. Versuch einer psychologischen Theorie der akquirierten Hysterie, vieler Phobien und Zwangsvorstellungen und gewisser halluzinatorischer Psychosen," *Neurologisches Zentralblatt*, Nos. 10 and 11.

GUTTMANN, G. (1968). Evoziertes Potential und Erleben. In: M. Irle (Ed.). *Bericht des 26. Kongresses der Deutschen Gesellschaft für Psychologie*, Göttingen: Hogrefe, pp. 189–195.

GUTTMANN, G. and BAUER, H. (1982). Learning and information processing in dependence on cortical DC potentials. In: R. Sinz and H. Rosenzweig (Eds.). *Psychophysiology 1980*, Jena: Fischer & Elsevier.

GUTTMANN, G. and BAUER, H. (1984). The Brain Trigger Design: A Powerful Tool to Investigate Brain-Behavior Relations. In: R. Karrer, J. Cohen and P. Tueting (Eds.) *Brain and Information: Event Related Potentials. Annals of the New York Academy of Sciences*, vol. 425, New York: New York Academy of Sciences, pp. 671–675.

GUTTMANN, G. (1992). Zur Psychophysiologie des Bewußtseins. In: G. Guttmann and G. Langer (Eds.) *Das Bewußtsein. Multidimensionale Entwürfe*, Vienna: Springer.

GUTTMANN, G. (1995). Psychophysiology of hypnosis and other altered states of consciousness. In: *Hypnosis Connecting Disciplines*, Vienna: Med. Pharm. Verlagsgesellschaft.

PRIBRAM, K. H. (1971). *Languages of the brain. Experimental paradoxes and principles in neuropsychology*, Englewood Cliffs, NJ: Prentice Hall.

ROHRACHER, H. (1935). "Die gehirnelektrischen Erscheinungen bei geistiger Arbeit," *Zeitschrift für Psychologie*, 136, 308–324.

ROHRACHER, H. (1937a). "Die gehirnelektrischen Erscheinungen bei Sinnesreizen," *Zeitschrift für Psychologie*, 140, 274–308.

Rohracher, H. (1937b). "Die gehirnelektrischen Erscheinungen bei verschiedenen psychischen Vorgängen," *Pontificia Academia Scientiarum "Commentationes"*, 1, 89–133.

Rohracher, H. (1940). "Die elektrischen Vorgänge im menschlichen Gehirn," *Zeitschrift für Psychologie*, 149, 209–279.

Vitouch, O., Bauer, H., Gittler, G., Leodolter, M. and Leodolter, U. (1997). "Cortical activity of good and poor spatial test performers during spatial and verbal processing studied with Slow Potential Topography," *International Journal of Psychophysiology*, 27, 183–199.

Walter, W. G., Cooper, R, Aldridge, V. J., McCallum, W. C. and Winter, A. L. (1964). "Contingent negative variation: an electric sign of sensorimotor association and expectancy," *Nature*, 203, 380–384.

Erklärung der Abbildungen.

Tafel II.

Fig. 1. Spinalganglion, Goldpräparat.

Gez. bei Hartnack $3/8$. Vergrösserung 435. Mehrere breite durchziehende Fasern, einige mit Fasertheilung.

$h\ W$ = hintere Wurzel.
dz = breite durchziehende Faser.
dz' = durchziehende Faser.
zf = Zellenfaser.
Th = Theilungen von Fasern.
Th' = Theilung einer breiten Faser in zwei ungleich starke Äste.
ang = angelehnte Faser.

Fig. 2. Rückenmark von Petromyzon marinus. Ansicht von der vorderen Fläche. Alkohol-Karminpräparat. Vergrösserung 115.

Vordere oberflächliche Faserkreuzung.

C = Centralcanal.
Mf = die Müller'schen (kolossalen) Fasern.
Vhz = Vorderhornzellen.
Cmf = Vordere Faserkreuzung.
Th = Theilungen von Fasern.

Fig. 3. Eine hintere Wurzel mit oberflächlicher Hinterzelle auf der Pia mater. Alkohol-Karminpräparat. Vergrösserung 220.

$h\ W$ = hintere Wurzel.
zf = Zellenfaser.
ohz = oberflächliche Hinterzelle.
$auf.f$ = aufsteigende Faser.

Fig. 4. Vordere Wurzel, Goldpräparat. Vergrösserung 285.

$v\ W$ = vordere Wurzel.
d = dorsaler Ast.
v = ventraler Ast.
kz = kleine, eingelagerte Zelle.

Harald Leupold-Löwenthal

Freud as a Neurologist

I was able to localize the site of a lesion in the medulla oblongata so accurately that the pathological anatomist had no further information to add; I was the first person in Vienna to send a case for autopsy with a diagnosis of polyneuritis acuta.[1]

In these words Sigmund Freud described his activity and his success as a clinical neurologist at the Vienna General Hospital in 1886 in his *Selbstdarstellung* [*An Autobiographical Study*] of 1925. The case study based on patient data from the Fourth Medical Department headed by Franz Scholz was published in 1886 in the *Wiener Medizinische Wochenschrift* under the title of *Akute multiple Neuritis der spinalen und Hirnnerven* [*Acute multiple neuritis of the spinal and cranial nerves*] as Freud's first scientific paper after being appointed *Privatdozent*. Freud justified the publication of the study as follows: "*Since according to the pathologist such a case has not previously been autopsied in Vienna, I permit myself to relate the case history here.*"[2]

This work is cited in the *Inhaltsangaben der wissenschaftlichen Arbeiten des Privatdocenten Dr. Sigm. Freud (1877–1897)* [*Abstracts of the Scientific Writings of Dr. Sigm. Freud (1877–97)*], which he submitted to the Council of the Faculty of Medicine when he applied for an appointment as 'Professor Extraordinarius'. He ends the abstract by saying: "*According to the pathologist [Kundrat], this was the first post-mortem finding of polyneuritis to be made in Vienna.*"[3]

Ten and indeed even forty years after his description of this neurological case, Freud still emphasized the priority of clinical diagnosis. In *An Autobiographical Study* he goes on to say that the fame of his diagnoses and their post-mortem confirmation had brought him an influx of American physicians to whom he lectured about the patients in his department "*in a sort of pidgin-English*".[4] But he added: "*About the neuroses I understood nothing.*" When on one occasion he introduced to his audience a

---

[1] Freud, S. (1959). An Autobiographical Study. In: *Standard Edition,* vol. XX, London: Hogarth Press, p. 12.

[2] *Wiener Medizinische Wochenschrift.* No. 6, 1886, p. 167.

[3] Freud, S. (1962). Abstracts of the Scientific Writings of Dr. Sigmund Freud (1877–97). In: *Standard Edition,* vol. III, London: Hogarth Press, p. 236.

[4] Freud, S. (1959). An Autobiographical Study. In: *Standard Edition,* vol. XX, London: Hogarth Press, p. 12.

neurotic suffering from a persistent headache as a case of "chronic localized meningitis", they rose in revolt against him and his course came to an end.[5] *"By way of excuse I may add that this happened at a time when greater authorities than myself in Vienna were in the habit of diagnosing neurasthenia as cerebral tumour."*[6]

Freud had spent most of his medical studies in the laboratory (with Claus, von Brücke, Stricker and Meynert). *" The various branches of medicine proper, apart from psychiatry, had no attraction for me."*[7]

In his evaluation of Freud's *Habilitation* (21 January 1885), Brücke wrote that of all the scientific papers submitted by Freud only the earliest one made no reference to the nervous system. It was the study *Beobachtungen über Gestaltung und feineren Bau der als Hoden beschriebenen Lappenorgane des Aals* [*Observations on the configuration and finer structure of the lobed organs in eels described as testes*], which the third-year student was allowed to publish in volume LXXV of the *Sitzungsberichte der kaiserlichen Akademie der Wissenschaften* in 1877 – an honor granted to only a few students. After switching to Brücke's Institute, the evaluation continues, Freud published „... further tissue-related studies on the nervous system as well as studies on the methods of its anatomical representation".

Freud did not receive his medical degree until 1881. In 1882, on Brücke's advice he abandoned his theoretical career and turned to the clinical subjects instead. His general medical training had been quite scanty, however, and would not have permitted him to set up a general practice. He was thus forced to turn to neurology which was also natural given the neurohistological and neuroanatomical research which he had carried out e.g. at Meynert's Psychiatric Clinic. This may have been one reason why Freud worked for the relatively extended period of 14 months at the neurological ward of the Fourth Medical Department of the General Hospital headed by Scholz, whom he did not hold in high esteem and with whom he did not get along well. The result was that the young *Privatdozent* had no patients for his lectures. When he joined the public Institute for Children's Diseases run by Max Kassowitz and "was put in charge of neurology"[8], Freud had many patients but the faculty allowed neither him nor the other *Dozenten* to give lectures at the Kassowitz Institute. The Kassowitz Institute was actually a free outpatient clinic for children which now that Freud was associated with it offered also neurological examinations. The number of patients there rose from 1,882 in 1872 to 8,541 in 1888, and in 1902 (after Freud had already left the Institute) to 21,600 patients.

---

[5] Ibid.
[6] Ibid.
[7] Ibid., p. 10
[8] Ellenberger, H. F. (1970). *The Discovery of the Unconscious. The History and Evolution of Dynamic Psychiatry*, New York, p. 476.

After his return from Paris Freud set up his medical practice as a specialist for nervous diseases. He did not, however, – as is often intimated – devote his efforts merely to the study of hysteria which, incidentally, was not just motivated by his research interests, but also by practical necessity: *"Anyone who wants to make a living from the treatment of nervous patients must clearly be able to do something to help them. My therapeutic arsenal contained only two weapons, electrotherapy and hypnotism, for prescribing a visit to a hydropathic establishment after a single consultation was an inadequate source of income."*[9]

At the same time Freud continued to work as a neurologist, volunteering his services at the Kassowitz Institute, and published a number of significant neurological papers. In 1888 another case study appeared: *Über Hemianopsie im frühesten Kindesalter* [*On hemianopsia in earliest childhood*].[10] He reported on two young children with semilateral dysopia as part of an infantile hemiplegia. According to Reinhard's and Wilbrand's comments known at the time it seems to have been the first reported observation of this symptom. In 1891 this was followed by the monograph *Klinische Studie über die halbseitige Cerebrallähmung der Kinder* [*Clinical Study of the Unilateral Cerebral Palsies of Children*][11], written jointly with Oskar Rie and published in the *Beiträge zur Kinderheilkunde* edited by Dr. M. Kassowitz. The authors described a "choreatic paresis" which from the outset exhibited a hemichorea rather than hemiplegia. They pointed out the close connection between epilepsy and cerebral palsies of children "*in consequence of which some cases of apparent epilepsy might deserve to be described as 'cerebral palsy without palsy'*".[12]

The authors also argued against Strümpell's hypothesis on the role of *polioencephalitis acuta* as constituting the anatomical basis of unilateral cerebral palsy. In *Neurologisches Centralblatt*, No. 12, 1893, Freud reported *Über ein Symptom, das häufig die Enuresis nocturna der Kinder begleitet* [*On a symptom which often accompanies enuresis nocturna in children*].[13] This study reported on the observation of a "*hypertonia of the lower extremities the significance and implications of which are unexplained.*"[14]

There was not yet any psychological explanation even though the "Vorläufige Mitteilung" ["preliminary communication in collaboration with Dr. J. Breuer"]

---

[9] Ibid., p. 39 f.

[10] *Wiener Medizinische Wochenschrift* (1888). Nos. 32 and 33 [English summary in *Standard Edition*, vol. III, p. 239].

[11] Freud, S., Rie, O. (1891). *Klinische Studie über die halbseitige Cerebrallähmung der Kinder,* Vienna [English summary in *Standard Edition*, vol. III, pp. 241–242].

[12] Standard Edition, vol. III, p. 242

[13] *Neurologisches Centralblatt*, (1893). No. 12, pp. 735–737 [English summary in *Standard Edition*, vol. III, p. 243].

[14] Ibid.

Über den psychischen Mechanismus hysterischer Phänomene [*On the psychical mechanism of hysterical phenomena*][15] appeared at the beginning of 1893 in the first issue of *Neurologisches Centralblatt*. In 1895, in vol. 14 of the same journal, we find Freud's last clinical-neurological publication on *meralgia paraesthetica, Über die Bernhardt'sche Sensibilitätstörung am Oberschenkel* [*On Bernhardt's disturbance of sensibility in the thigh*]. In Freud's own words: "*A self-observation of this harmless affection, which is probably traceable to local neuritis; and a report of some other cases, including bilateral ones.*"[16] In his letters to Fliess, Freud referred to this and other papers contemptuously as "*Schmarren*" ("a rubbishy trifle") – which, of course, raises the question why they were published in the first place.

In 1895 Hermann Nothnagel asked Freud to write an article for his encyclopedia *Specielle Pathologie und Therapie* and by 1897 Freud had written *Die infantile Cerebrallähmung* [*Infantile Cerebral Palsies*] for this reference work.[17] Freud himself did not think very highly of this text. In several letters addressed to Fliess, he repeatedly mentioned the arduous duty of having to write this contribution. In the article for Nothnagel's encyclopedia the extensive (170 pages!) study of 1893 *Zur Kenntnis der cerebralen Diplegien des Kindesalters (im Anschluß an die Little'sche Krankheit)* [*An Account of the Cerebral Diplegias of Childhood (in Connection with Little's Disease)*][18] was summarized and supplemented together with the above-mentioned publication that Freud had written jointly with Oskar Rie.

The *Cerebral Diplegias* may be regarded as a standard work. In an article on "Freuds Leistungen auf dem Gebiet der Neurologie"[19] the psychoanalyst Rudolf Brun, a former assistant of von Monakow's, assessed this study as being "the most thorough and most complete that has been written on cerebral diplegias to date (1936)."

As late as 1939 – a time when Jewish authors were looked down on and could no longer be cited – Johannes Lang referred to Freud's studies in the 5th volume, *Krankheiten des Nervensystems*, of the *Handbuch der Inneren Medizin*, edited by Staehelin and von Bergmann.[20] In his introduction to *Cerebrale Diplegien des Kindesalters* Freud included some reflections on his clinical-empirical approach which demonstrate a scientific attitude in clinical work that is also found in his psychoanalytic research: "*Whenever a type of illness is to be designated by the same name as a symptom thereof, mis-*

---

[15] *Neurologisches Centralblatt*, vol. 12 (1893), pp. 4–10, 43–47. [English translation in *Standard Edition*, vol. II, pp. 3–17; summary in *Standard Edition*, vol. III, p. 244].

[16] *Neurologisches Centralblatt* (1895). vol. 14, pp. 491; summary in G. W. vol. I, p. 485 [English summary in *Standard Edition*, vol. III, p. 253].

[17] Summary in *G. W.*, vol. I, pp. 487 f. [English summary in *Standard Edition*, vol. III, p. 256].

[18] *Beiträge zur Kinderheilkunde*, No. III, ed. by Dr. M. Kassowitz [English summary in *Standard Edition*, vol. III, p. 245].

[19] *Schweiz. Archiv Neurologie und Psychiatrie*, 37, 9, 183–200.

[20] p. 569.

understandings can arise. I would thus like to stress from the outset that this study does not deal with the symptom of cerebral diplegia but with the states of disease which for the time being, before more is known about them, are named after their most striking symptom."[21] And the accompanying footnote says: *"If I proceed in this way, I am openly acknowledging a usage that all other contemporary authors have equally accepted."*[22]

A further remark reveals his careful methodology: *"At the beginning of this study I cannot say what the processes of the disease are, which I have summarized here as cerebral diplegia. I would be very satisfied if this study would result in such an explanation. As always, when a definition is lacking due to our insufficient knowledge, of the essential features, I have to rely on a kind of personal identity, on the agreement of my cases with those regarded as models of cerebral diplegia."*[23]

In his textbook on nervous diseases published in 1913, Oppenheim began the chapter on "Cerebral Palsies in Children" with the article which Freud had written for Nothnagel's encyclopedia.[24] As a specialist in nervous disorders, Freud was at the cutting edge of his profession at the time: As a neurologist he tried to classify symptoms first in terms of syndromes and then in terms of diseases. As a neuropathologist he tried to determine the localization of lesions corresponding to the clinical phenomena, but no longer by means of neuroanatomical or neurohistological methods which he was certainly capable of, since there were neither autopsy nor laboratory facilities at the children's outpatient clinic.

In *An Autobiographical Study*[25], Freud explained that he had published a number of larger studies on unilateral and bilateral cerebral palsies in children while working at the Kassowitz Institute in the years after his return from Paris in 1886. However, a few pages later he states that: *"During the period from 1886 to 1891, I did little scientific work, and published scarcely anything."*[26]

In 1891, the work that Freud had written in collaboration with Oskar Rie on the unilateral cerebral palsies in children appeared. By then Freud had also published two neurohistological studies, *Über die Beziehung des Strickkörpers zum Hinterstrang und Hinterstrangskern, nebst Bemerkungen über zwei Felder der Oblongata* [*On the relation of the restiform body to the posterior column and its nucleus with some remarks on two fields of the medulla oblongata*][27] (begun with Darkshevich in Paris)

---

[21] *Beiträge zur Kinderheilkunde* (1893). p. 2.
[22] Ibid.
[23] Ibid.
[24] H. Oppenheim (1913). *Lehrbuch der Nervenheilkunde für Ärzte und Studierende*, vol. II, Berlin, p. 1093.
[25] In: *Standard Edition*, vol. XX, London: Hogarth Press, p. 14.
[26] Ibid., p. 19.
[27] Jointly with Dr. L. Darkshevich from Moskow; *Neurologisches Centralblatt* (1886). No. 6, see G. W. vol. I, p. 470. [English summary in: *Standard Edition*, vol. III, London: Hogarth Press, p. 237].

and *Über den Ursprung des Nervus acusticus* [*On the origin of the auditory nerve*][28], as well as two papers on clinical neurology, *Beobachtung einer hochgradigen Hemianästhesie bei einem hysterischen Manne* [*Observation of a severe case of hemi-anaesthesia in a hysterical male*][29] and *Über Hemianopsie im frühesten Kindesalter* [*On hemianopsia in earliest childhood*][30]. In 1887, Freud's *Bemerkungen über Cocainsucht und Cocainfurcht* [*Remarks on addiction to cocaine and the fear of cocaine*][31] had appeared — studies which Freud was perhaps not too eager to remember.

This period also witnessed the writing of the socalled *Entwurf einer Psychologie* [*Project for a Scientific Psychology*][32] and the collection of clinical material at the Kassowitz Institute as well as Freud's work with hysterical and neurotic patients with hypnosis and pressure-point therapy. Between 1887 and 1891 Freud also wrote several articles for Villaret's *Handwörterbuch der gesamten Medizin*. Since the contributions were not signed and the article on the anatomy of the brain was shortened, Freud did not include them in his *Inhaltsangaben* [*Abstracts*] of 1897. He likewise refused to include *Zur Auffassung der Aphasien* [*On the Interpretation of Aphasias*], published in 1891, in his *Gesammelte Werke*, since it was neurological rather than psychoanalytical[33] even though it was a study which he held in high esteem.

Perhaps Freud was thinking of this period when he began his account of the epicrisis of the case of "Fräulein Elisabeth von R." in his *Studien über Hysterie* [*Studies on Hysteria*] with the following words: *"I have not always been a psychotherapist. Like other neuropathologists, I was trained to employ local diagnoses and electro-prognosis..."*[34]

In this context it is also interesting to take a look at the themes of Freud's lectures as a *Dozent*. From the summer semester of 1886 to the summer semester of

---

[28] *Monatssch. für Ohrenheilkunde* (1886). vol. 20, summarized in *G. W.*, vol. I, pp. 470–471 [English summary in: *Standard Edition*, vol. III, London: Hogarth Press, p. 238].

[29] Lecture held at the *k. Gesellschaft der Ärzte* (Society of Physicians) on Nov. 26, 1886, and published in *Wiener medizinische Wochenschrift* (1886). Nos. 49 and 50. [English summary in: *Standard Edition*, vol. London: Hogarth Press, III, p. 238].

[30] *Wiener medizinische Wochenschrift* (1888) vol. 38, summarized in *G. W.*, vol. I, pp. 461–488. [English summary in: *Standard Edition*, vol. III, London: Hogarth Press, p. 239].

[31] *Wiener medizinische Wochenschrift* (1887). vol. 37, sumarized in *G. W.*, vol. I, pp. 461–488. [English summary in: *Standard Edition*, vol. III, London: Hogarth Press, p. 239].

[32] Aus den Anfängen der Psychoanalyse. Briefe an Wilhelm Fliess, Abhandlungen und Notizen aus den Jahren 1887–1902, ed. by M. Bonaparte, A. Freud and E. Kris. (1950). *Imago*, London. [English translation in: *Standard Edition*, vol. I, London: Hogarth Press, pp. 177–387 (excerpts only)].

[33] A shortened English version is, however, included in Freud, S. (1957). The Unconscious. In: *Standard Edition*, vol. XIV, London: Hogarth Press, pp. 206–215 [Translator's note].

[34] Freud, S. (1955). Case Histories. In: *Standard Edition*, vol. II, London: Hogarth Press, p. 160.

1892, all the lectures he gave were neuropathological and clinical-neurological in nature. In the winter semester of 1892/93, the theme of "The Theory of Hysteria" came up for the first time, but did not figure again until the summer semester of 1894. Until that time Freud had thus lectured on neuropathology and neurology. In her perceptive article in *Urbuch der Psychoanalyse. Hundert Jahre Studien über Hysterie*[35], Ilse Grubrich-Simitis describes Freud's and Breuer's invention of the art of listening. Since the disorders of their patients were related to the function of speech, "the usual precise study of the symptoms practically enforced a new manner of auditive attention".

Grubrich-Simitis thus alludes to an observant clinical attitude which, as she correctly notes, was characteristic of both Breuer and Freud. In addition to reflecting a certain fundamental scientific attitude, this approach was also the product of a shared experience: medical training and working in the context of what is referred to as the Second Vienna Medical School, which turned fin-de-siecle Vienna into a world-famous center of medicine with Rokitansky and Skoda as the first great teachers after von Feuchtersleben had implemented Türckheim's reform of the study of medicine. They were succeeded by von Brücke, Nothnagel, Billroth, Meynert and their students, one of whom was Freud.

As a pathological anatomist and internist, Rokitansky had held the view that empirical medicine should be replaced by the scientific analysis of the findings based on autopsy and microscopic examinations. He promoted neurology; in 1846, he established a neurological ward at the General Hospital in Vienna for Ludwig Türk, whose *Abhandlung über Spinalirritation* had already met with international acclaim. Türk then developed a method for examining secondary degenerations as a means of studying the structure of the central nervous system. He located the pyramidal tracts and introduced the theory of systemic neurological disorders. After his death the neurological ward was closed, with the exception of the outpatient clinic, which was then headed by Nathan Weiss and after the latter's suicide[36] by Moriz Rosenthal until 1888.

Rokitansky had also induced Brücke and Meynert to come to Vienna. Their basic attitudes corresponded to his scientific view of a mechanistic and materialist monism. It was Brücke, who introduced modern laboratory medicine in Vienna. His assistants Sigmund Exner, Ernst von Fleischl-Marxow and Alexander Rollet were among the leading physiologists of their time. As a staunch anti-vitalist, Brücke was convinced that in biology and physiology observable and measurable

---

[35] Supplement to the limited facsimile edition, S. Fischer, 1995, p. 14.
[36] On this see Freud's letter to Martha Bernays of September 16, 1883. *Letters of Sigmund Freud: 1873–1939.* Ernst L. Freud (Ed.) (1970). London: Hogarth Press, p. 73.

physical and chemical forces rather than a *vis vitalis* had to be used as explanations. In 1842 Emil Du-Bois Reymond, who like Brücke was a student of Johannes Müller, formulated this as follows: *"Brücke and I pledged a solemn oath to put in power this truth: No other forces than the common physical chemical ones are active within the organism. In those cases which cannot at the time be explained by these forces one has either to find the specific way or form of their action by means of the physical mathematical method, or to assume new forces equal in dignity to the chemical physical forces inherent in matter, reducible to the force of attraction and repulsion."* [37]

In Ernst von Brücke's physiological laboratory Freud *"... found rest and full satisfaction – and men, too, whom I could respect and take as my models ..."* [38] Trying to live up to Rokitansky's ideals, the neuropsychiatrist Meynert sought "to give psychiatry the character of a scientific discipline by placing it on a solid anatomical basis."[39]

Freud worked in Meynert's laboratory, focussing particularly on anatomical studies of the brain which were Meynert's area of interest. Krafft-Ebing, too, followed a similar, only slightly modified approach, as he explained in his *Lehrbuch der Psychiatrie* in 1879: *"Madness is a disease of the brain"*[40]. He added, however, *"that in a certain number of autopsies of insane persons we fail to find palpable cerebral findings."*

Krafft-Ebing did not establish a pathologically and anatomically founded nosology. Instead, he adopted an aetiological-taxonomical approach which was complemented by clinical-functional classificatory principles. Considerable importance was attributed to clinical observation and clinical phenomenology. One only need recall his *Psychopathia sexualis* in this context. For Krafft-Ebing neuroses were to be regarded as indicating a healthy, "sprightly" brain, while psychoses indicated a "sickly" brain.

The internist Nothnagel was also intensely interested in neurology. He preferred physiological, i.e., functional diagnoses, attributing particular importance to clinical aspects. He taught above all clinical methodology and as an outstanding teacher instilled in his students and in physicians a sense of independent medical thinking and acting. In 1887 he had appointed Frankl von Hochwarth head of the neurological outpatient clinic of his ward for internal medicine. Other important neurologists of this school included Wagner-Jauregg, Obersteiner and von Economo, to name but a few. In this cursory sketch I have tried to convey an impression

---

[37] Bernfeld, S. (1944). "Freud's Earliest Theories and the School of Helmholtz," *The Psychoanalytic Quarterly,* 13, 348. [German translation: Bernfeld, S. and Cassirer-Bernfeld, S. (1988). *Bausteine der Freud-Biographik.* Ed., trans. and introd. by Ilse Grubrich-Simitis, Frankfurt/Main, pp. 62 f].

[38] Freud, S. (1959). An Autobiographical Study. In: *Standard Edition,* vol. XX, London: Hogarth Press, p. 9.

[39] Meynert, T. (1884). *Psychiatrie. Klinik der Erkrankungen des Vorderhirns,* Vienna, p. VI.

[40] Krafft-Ebing, R. von (1879). *Lehrbuch der Psychiatrie auf klinischer Grundlage für Praktische Ärzte und Studierende,* Stuttgart, vol. 2, pp. 56.

of the medical environment in which Sigmund Freud, a "reluctant physician", received his formative scientific and medical impressions.

In her book Erna Lesky speaks of the *"circle of neurologists, busy with nosography, microscopy and electricity…"*[41], which Freud entered in 1882 and of which he was certainly a part. Lesky also writes that his life's work extended far beyond the 19th century, while his basic concepts were still rooted in that century.

A first change in Freud's attitude took place in Paris under the influence of Charcot, who opened up new possibilities for him: to defend *"… the rights of purely clinical work, which consists in seeing and ordering things against the encroachments of theoretical medicine,"*[42] and to learn that clinical facts can also provide material for scientific work and even take priority over theory. Freud's turning away from brain research and neurology towards psychology began in Paris, but not without a sometimes serious emotional crisis which had to be overcome as I described in 1986 in *A Viennese in Paris*[43]. Freud had left neuroanatomy behind, even though he could perhaps have continued some research alongside his private practice as a specialist in nervous diseases. He had *"in a sense* [found] *a new professional (transitional) identity which was autonomous and more in keeping with reality and which permitted him to fulfill old wishes in a satisfying way."*

In Charcot, Freud had found a teacher who allowed him to identify with him and idealize him, without, however, entirely relinquishing his critical distance. This man had been *"a poor devil"* (as Freud also saw himself), but had attained fame, success and influence and was an important scholar to boot. He did not have Brücke's *"terrible eyes"* or Nothnagel's straightforwardness or Meynert's unpredictable ways – all of whom had been among Freud's teachers, patrons and promoters.

In *Zur Geschichte der psychoanalytischen Bewegung* [*On the History of the Psychoanalytic Movement*] Freud wrote in 1914: *"I myself had only unwillingly taken up the profession of medicine, but I had at that time a strong motive for helping people suffering from nervous affections or at least for wishing to understand something about their states."*[44] It is no coincidence, I believe, that this work is preceded by the motto on the coat of arms of the City of Paris: *Fluctuat nec mergitur*. After returning from Paris, Freud worked intensively in the clinical-neurological field at the Kassowitz

---

[41] Lesky, E. (1976). *The Vienna Medical School of the 19th century [Die Wiener medizinische Schule im 19. Jahrhundert]*, Trans. from the German by L. Williams and I. S. Levij, Baltimore and London: Johns Hopkins Univ. Press, p. 355.

[42] Freud, S. (1962). Charcot obituary note. In: *Standard Edition*, vol. III, London: Hogarth Press, p. 13.

[43] Leupold-Löwenthal, H. (1997). *Ein unmöglicher Beruf. Über die schöne Kunst ein Analytiker zu sein*, Vienna, p. 229.

[44] Freud, S. (1957). On the History of the Psycho-Analytic Movement. In: *Standard Edition*, vol. XIV, London: Hogarth Press, p. 9.

Institute and began using the methods of suggestion and hypnosis to study and treat hysteria and neuroses in his private practice. He gradually withdrew from the neurological clinic and as is evidenced by his publications and the content of his lecture course, it is only from 1895 on that his research interests were purely psychological. Although he continued to assume a biological basis for psychological processes, claiming e.g. that the instincts had a biological basis, his own method of psychoanalysis, which he regarded as a research method, is really psychological.

In his *Entwurf* [*Project*], the manuscript that remained unknown until the 1950s, he tried to develop a "psychology on a neurological basis". I would like to point out that the English translation of the title, namely *Project*, gives a programmatic twist to the *Entwurf* [45] that was never intended (neither by the editors who chose the title nor by Freud, who sent the manuscript to Fliess). In his letters to Fliess, Freud had expressed his astonishment about his mental and psychological state while writing the text.

The *Entwurf* was still written in the spirit of his teachers, in keeping with the basic outlook of a biological (mechanistic) monism. In his study *Zur Auffassung der Aphasie* [*On Aphasia*], which Freud held in high esteem and described as "speculative", he left the approach based on neuronal and localization theories entirely behind and clearly embraced the position of a psychophysical parallelism which he retained more or less unchanged until his last work, the *Abriß der Psychoanalyse* [*An Outline of Psycho-Analysis*].

The "psychical apparatus" later described by him is no longer the central nervous system, but a construct allowing him to link psychological hypotheses with the clinical empiricism of the psychoanalytic process. The therapeutic effect of psychoanalysis as psychotherapy was always of secondary importance to him. In the first instance, it was *"a method for studying the unconscious"*. Freud may have had a remark in mind that had been made by Hofrat Meynert in a discussion that took place during a meeting of the Society of Physicians in June 1889, where Meynert postulated that *"a successful treatment proves nothing, it must first be proven itself"*.

Behind the late 19th century well-cushioned, embroidered and decorated easy chair lurks the "project of modernism", which can be seen everywhere in what Walter Benjamin, in his posthumous writings, has called "Etui"- or "Futteraldenken".

It is not in spite of the fact, but precisely because Sigmund Freud was a child of his time, while at the same time trying to resist and overcome obfuscation, that he and psychoanalysis were of such crucial importance for the "project of modernism".

---

[45] See footnote 32. The German title "Entwurf" was chosen by the editors! [And the English title was chosen by Freud's translator James Strachey – Translator's note.]

Erklärung der Abbildungen.

Tafel III.

Fig. 1. Flächenschnitt der Pia mater mit 5 hinteren Wurzeln und den oberflächlichen Fasern und Hinterzellen. Chromsäurepräparat mit Gold gefärbt. Vergrösserung 50. Bei hW^2, hW und hW' zwei halbe Wurzeln an Stelle einer einzigen.

$hw^1 - hw^5 =$ hintere Wurzeln.
$ohz =$ oberflächliche Hinterzellen.
$auf. f =$ lange aufsteigende Fasern.
$qhz =$ in die Wurzel eingelagerte Hinterzelle.
$FZ =$ Faserzusammensetzungen im Verlaufe der aufsteigenden Fasern.

Fig. 2. Flächenschnitt (Frontalschnitt) durch die Pia mater und die umgebenden Gewebe. Chromsäurepräparat mit Gold gefärbt. Vergrösserung 105.

Zellen in den queren Verlauf der Wurzel eingelagert bei qhz und qhz'.

$sk\ G =$ sog. skeletbildendes Gewebe um den Rückenmarkskanal.
$D =$ Dura mater.
$Ar =$ Arachnoidealraum.
$Spg =$ Spinalganglion.
$G =$ Gefässdurchschnitt.
$M =$ Muskel.
$hW =$ hintere Wurzel.
$U =$ Umbiegung der hinteren Wurzelfasern im Rückenmark.
$vW =$ vordere Wurzel.
$Gf =$ Gefäss.
$qhz =$ querliegende, in die Wurzel eingelagerte Hinterzellen.

Fig. 3. Hintere Wurzel mit aufsteigenden Fasern und oberflächlicher Hinterzelle aus einem Flächenschnitt der Pia mater. Chromsäure-Goldpräparat. Vergrösserung 110.

$hW =$ hintere Wurzel.
$auf. f =$ aufsteigende Fasern aus einer früheren Wurzel.
$ohz =$ oberflächliche Hinterzelle.

Fig. 4. Isolirte oberflächliche Hinterzelle auf der Pia mater. Chromsäure-Goldpräparat. Vergrösserung 110.

$ohz =$ oberflächliche Hinterzelle.
$zf =$ Wurzelfortsatz derselben.
$hw =$ Umbiegung desselben zur hinteren Wurzel.
$c =$ centraler Fortsatz.

Malcolm Pines

Neurological Models and Psychoanalysis

In this communication I shall draw out some of the underlying features of the neurological model used by Freud, notably in his *Project for a Scientific Psychology* (1895) with the very different neurological models used by Kurt Goldstein and Paul Schilder. Goldstein was born in 1878 and graduated in medicine in 1903, Schilder, who was eight years his junior, graduated in medicine in 1909 and went on to take an additional doctorate in philosophy from the University of Vienna during World War One. Both Goldstein and Schilder had strong philosophical interests which they integrated with their models of the human being; both were influenced by Husserl in philosophy and by Wernicke in psychiatry. Goldstein and Schilder covered very wide areas in their clinical work and publications and it is through their advocacy of a holistic organismic approach that they should both be remembered and celebrated, though their example and influence is nowadays not widely recognised. Goldstein is best known through his studies of the brain-injured patient in which he showed how the basic deficit in brain damage is loss of the abstract attitude and its replacement by the concrete. Later he applied this work to schizophrenia. Schilder's work culminated in his *Image and Appearance of the Human Body* published in 1935.

After having sketched out the distinctions between the models of Freud, Goldstein and Schilder, I shall also briefly relate to developments in the understanding of language and communication which seem to follow the same developmental path: in particular I shall briefly refer to Saussure, Vigotsky and Bakhtin.

Freud's *Project* (1895), preceded by his monograph *On Aphasia* (1891), represented his most ambitious attempt to create a neurodynamic view of the psyche. He used a neuronic model, the fundamental unit of which is the reflex arc; the model is concerned with drive regulation. In a letter to Fliess he wrote, "*Everything fell into place, the cogs meshed, the thing really seemed to be a machine which in a moment would run of itself*". (*Origins of Psychoanalysis*, Letter 32, page 129). Freud's neurones were theoretical neurones, his neurological experience having preceded Cajal's discovery of the anatomical neurone – and he built them into a complex network. Freud's concern was with somatic energy, with the structure that contained and discharged this energy. This mental apparatus developed in response to external stimulation, for the organism is inevitably dependent on the objective world for "the exigencies of life". The basic problem of the mental apparatus is to cope with bodily needs and to regulate drive energy. The capacity for distinguishing memory and perception, for delaying discharge, for developing the ego pathways is elaborated. Walter Stewart writes of, "the beauty of this self-regulation, self-reporting control of energy flow in the neuronic system".

Roger Smith in his wide ranging study of the Concept of Inhibition in 19th century culture and science, writes of the "language of ordered relations that acquired authority in the 19th century" (page 222), and how, in Freud's work, we can see that the order that makes society possible becomes the order within the person through which power is distributed and differentiated. Freud's six assumptions were causal determinism, explanation by dynamic forces, an economic model of energy and a structural, developmental and hierarchical approach to functional interaction. This cluster of assumptions, Freud's frame of reference, concerned the dynamic interaction between different elements of nervous or psychic energy and the way in which their interaction was ordered "in spatial, developmental and evolutionary terms". Freud's metaphors related to a mental apparatus and the forces operating within it, these metaphors shifting from spatial to dynamic, but even so these economic and dynamic psychic metaphors related to 19th century economic concepts of limited resources, of inhibition resulting from the clash of forces, emphasising the values of order and control.

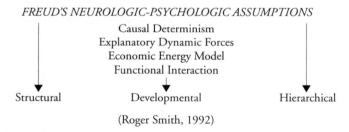

(Roger Smith, 1992)

What is it that so radically differentiates the later models used by Goldstein and Schilder from this 19th century model of Freud's? It is the organismic and holistic position in which organism and environment cannot be studied in isolation, for together they are the fundamental units for study. To quote Schilder, "A body without a world is just as unthinkable as a world without bodies". The knowledge of our corporality, the three-dimensional concept of our self which we carry in us must also be the knowledge of the outside world. Schilder's concept of the body image incorporated Freud's view of the ego as primary body ego, the postural model of the British neurologist Henry Head and Wernicke's somatopsyche. And society is incorporated in the body because the organism has to take its place in the world. There is no possibility of studying or understanding the body or the mind without taking into account circumstances and context.

Walter Riese, a distinguished neurologist who was also sympathetic to psychoanalysis, has completed Freud's psychoanalytic and neurologic models as they appear in his early writings.

NEUROLOGY	PSYCHOANALYSIS
Associationism	Topographic model
	Structural model
Aphasia as interrupted associations	Conversion hysteria
	part isolated not entering into association with ideas
	making ego take 'wrong path'
	Subconscious affective associations
'Central Apparatus of speech'	'Psychic Apparatus'

(Walter Riese, 1958)

Freud's model of aphasia was based on associationism, the then prevalent model in neurology and psychology. Thus the speech disorder derived from interrupted associations in the cortex. It is interesting to note that Freud wrote of the "central apparatus of speech" which, as Ervine Stengel put it, represented the elder brother of the "psychic apparatus". Walter Riese writes that Freud "failed to do justice to the new and creative factor implied in human language". By this he means that there is a qualitative conceptual leap from this 19th century associationist anatomical-physiological model of speech to the concept of the "animal symbolicum", to use the term of Ernst Cassirer (incidentally Kurt Goldstein's cousin). Cassirer describes man as not living in a merely physical universe but in a symbolic one employing linguistic forms, artistic images, medical symbols or religious rites. Following this direction in 1926 the great British neurologist Sir Henry Head, defined aphasia as "disorders of symbolic thought and expression". Kurt Goldstein is described by Walter Riese as "thoroughly humanistic though experimental at the same time, a shining example of the compatibility of both these perspectives". Goldstein, clearly influenced by his cousin Cassirer, often refers to the great Wilhelm von Humboldt, who taught that language is the organ of thought and that intellect and speech are one and inseparable. Goldstein, Head and Schilder occupy the conceptual domain in which understanding language, its development and its disorders, is inseparable from studying the human being as a communicative animal. This social animal engages in meaningful communication with himself and with his fellows, because he is a listener to himself and thereby knows that the other will understand his utterance. The human being as an articulate communicator is inseparable from the formative cultural matrix of language and behaviour.

COMPARISON OF FREUD'S MODEL AND THE GOLDSTEIN-SCHILDER MODELS

	Freud	Goldstein – Schilder
Field of Observation	Individual CNS	Organism
Environment	Source of Stimuli	Intrinsic: Figure-ground
Concepts	Tension Reduction	Active Adaptation
	Biological Drives	Communication
	Association Psychology	Gestalt Psychology
Neurological Model	Evolutionary	Neuronal Network
	Doctrine of Levels	
Clinical	Aphasia in Children	Brain Damaged Adults
		Schizophrenia

Goldstein, a phenomenologist, differentiated himself from psychoanalysis, but Schilder was invited by Freud to join the Vienna Psychoanalytic Society though he never underwent personal analysis. Both Goldstein and Schilder grew up in the new psychology of Gestalt and both were opposed to Freud in what they saw as the psychoanalytic project, which is to assemble the whole organism from a study of its parts, for example the structural model, whereas their starting point was the organism in its totality, however early or primitive in form it might seem to be.

As Walter Riese (*The Reach of the Mind*, essays in memory of Kurt Goldstein, edited by Marianne Simmel, Springer Publishing Company New York) has written, "Goldstein insisted on the total unitary function of all nervous events, of all organic events". The Gestalt concept of figure and ground is always taken into account, a concept which as far as I know is absent in Freud. The driving forces of both normal and abnormal cerebral events are the actual situation in which the person finds himself, its demands, their relation to the possibilities remaining to the individual and his attempts to solve the problem set before him, even though they may overpower him and elicit catastrophic reactions. Goldstein rejected the reflex arc as a useful model of neuronal processes, for as a philosopher of biology he saw it as a product of the biological schema of the 19th century. His basic unit of study is the interdependence and reciprocity of parts.

Any disturbance to the central nervous system or to the organism itself produces changes in the functioning of the whole central nervous system and other physiological systems. Goldstein's study of language in Aphasia represents a conceptual leap forward from Freud's by his demonstration of the subtlety of disturbances of language that can only be elicited through the most careful attention to the way in which the sufferer approaches the task in hand. However, Goldstein shared Freud's opposition to doctrines of cerebral localisation as solely responsible for neurological symptoms and also shared Freud's acceptance of Hughlings Jackson's thesis of the hierarchy of levels in the central nervous system.

Amongst Goldstein's intellectual influences were the philosophers, Goldstein's cousin Ernst Cassirer, Edmund Husserl and Kant. My own analyst and teacher,

S H Fuchs was greatly influenced by Goldstein, having worked as his personal assistant for two years after World War One, prior to his training as a psychoanalyst.

Fuchs was also a junior contemporary and pupil of Paul Schilder, and acknowledged his genius. Fuchs greatly admired Schilder's work on the *Image and Appearance of the Human Body*, which had led Schilder to become a pioneer of group psychotherapy when he moved to America. Schilder's basic assumption was that the human organism intends to grasp and to achieve reality in life's most obvious senses as well as on its richest and deepest levels. In his basic frame of reference were *intentionality*, (Vigotsky) the synthetic function of the psyche (ego), the body image concept and the sociological situation of the human being. "The relation between outer world, body and self, is a fundamental human relation". He introduced the concept of the "sphere", influenced by Wernicke, this being the unformed background of experience from which object directed intentions emerge as reality-relevant thoughts and actions. He did not differentiate between conscious and unconscious in the way that psychoanalysis does; there is a continuum of consciousness within the sphere and Heinz Hartman acknowledged that Schilder's view of mental functioning anticipated concepts such as the autonomous functions of the ego.

Schilder insists that every sensation has a motor aspect, that perception involves activity and movement. The drives thrust outwards towards the world and other individuals have *ab initio* a full interest in the outside world and therefore in ist preservation; we have pleasure in activity and both our activity and our aggression express our interest in external objects. These views of Schilder anticipate modern investigations of infant-caregiver relations which do not support the concept of the phase of primary narcissism or early autism, a position which Schilder himself always took, disclaiming primary narcissism. Contemporary studies show how the infant actively investigates the presence of others and seeks from them the completion of his early activities. From the start all the senses are evoked and evolved – touch, taste, smell, balance, hearing and vision. For Schilder "The secret of life seemed to be the seeking after but never quite achieving closure". (Ziferstein). According to Schilder the synthetic function is a general characteristic of psychic life: the psyche manifests itself as a unit which represents itself in different aspects and builds itself up again and again in response to varying situations.

PAUL SCHILDER'S BASIC CONCEPTS
Intentionalism. Drive towards objects Sphere.
Unformed background of experience (unconscious)
Synthetic psychic function
Body image concept

Schilder's and Goldstein's insistence on studying the nervous system as a whole and on the organism's active interest in achieving a satisfying response is supported by modern neurophysiological research. F. Levin (*Mapping the Mind: the Intersection of Neuroscience and Mind*) describes the role of the general level of arousal of the brain. When the arousal level is below a certain threshold of excitement, the person's cortical activity seems to be limited to only one cortical association area at a time but when the threshold of interest is exceeded, the brain becomes activated as a whole and the various associative cortical parts come into communication with each other, as if in an internal conversation or dialogue. It is then that learning about the world can take place. Similarly Daniel Stern has shown through his research on infants that their interest in and openness to the world is at its maximum after feeding, after consummatory activity, for it is then that the infant, content and relaxed, is curious about and engages in rich interchanges with the world, primarily with the caregivers and it is here that the basis for the capacity for intimacy is laid down.

SCHILDER: THE BODY IMAGE

The picture of our own body which we form in our own mind, the way our body appears to ourselves is based on these constituents:
Physiological neurological
Libidinous
Sociological

Schilder's concept of the body image combines the physiological, the libidinal and the sociological. The physiological is investigated through disturbances in the body image, such as Gerstmann's syndrome, which he studied together with Gerstmann; the libidinous structure of the body image refers to developmental phases, the oral cannibalistic phase when the organism wishes to incorporate the world, the anal sadistic power struggle in relationship to the world which is turned into a masochistic submission in an attempt to incorporate that external power as part of the self, and eventually the erogenic.

LIBIDO AND BODY IMAGE

Narcissistic		
Oral cannibalistic		Bring body
Anal sadistic		close to
Genital		world
Grasping	Groping	Sucking
	(return with brain injury)	

A major feature of the body image is its sociology. Libido and society are inseparable; all body images are interconnected and are concerned with affectivity and closeness. Body images cannot exist in isolation, we demand that our body images be unified and insist on the close link between our bodies and those of others. The most solitary or withdrawn activities such as masturbation are social; the secrecy, the blushing, the fear that evidence of masturbation may be seen on the face demonstrate its social aspects. Changes in body image are always social and alter the body image of others and there is a constant exchange between these images. The total body image of others can be absorbed into our own and our own bodies can be projected as a whole into those of others. Movement, clothing, social activities such as dancing, all evoke changes in the body image.

CHANGES IN BODY IMAGE

Negative	Positive
Organic	Clothing
Psychotic	Adornment
Neurotic	Activity
Depersonalisation	

The body image is in some ways the sum of the body images of the community and is constituted according to the various relationships the community makes available. Concepts such as closeness, the near and the far, are impregnated with social experiences, for there are continuous interchanges between parts of our body images and those of others, through projection and appersonisation. Similarly, the ego cannot be constituted without others, being created in continual interchanges.

SOCIOLOGY OF BODY IMAGE

Grows beyond physical body into social space
Libido is social directed at other body images
All images are interconnected, affectively connected and continuously exchanged
Constructed reconstructed dissolved
Preservation of integrity of self and other – ethical imperative

This dance of the body images brings me to my last thesis, of the dialogue as the basic unit of human existence. We exist immersed in civil society, for which we require knowledge that is never completely in the head of any one of the individuals in its use. It is our culture that forms the invisible of our social activities – we are born into a world of sounds, languages, activities, from which we draw provisions and resources for forms of social behaviour, for becoming members of the community.

Dialogue is essentially taking part in joint action with others: we must respond by formulating appropriate utterances in reply to their utterances: we speak into a context that is not of our own making. And from this situation, that is speaking into a social context, moral knowledge emerges, for we depend on the judgements of others as to whether our expressions are ethically proper or not, a kind of knowledge we cannot have solely within ourselves. This is a form of knowledge, a knowing "from within" to contrast it with other kinds of knowledge such as knowing facts and "knowing how". To illustrate this, recently the group was joined by a patient who quickly seized the space-time of the group to speak amusingly about himself. The responses that he evoked, the ones that he took away with him and thought about, were of having been told firstly that he was not taking good care of himself by responding so immediately to intimate questions from others, such as details about his sexuality, and that he needed to be aware of this to safeguard his position both in the group and in society; and secondly, that though he was the sort of person who would frequently be invited to parties because he was so enlivening this is not the sort of person that people would want to offer employment to because they might have doubts of his reliability or capacity for work. He reflected on both these responses and felt that they had been indeed useful to him in enabling him to understand why despite all his activities he nevertheless remains friendless.

Here again infant caregiver research is beginning to show (Emde) that a morality of reciprocity and mutuality is encoded in early forms of behaviour, when sameness, difference, otherness, is experienced and is externalised and internalised in early forms of play. Through such experiences a sense of "we-ness", "us-ness" emerges, that Emde terms "the executive we". It is here that the powers of the world of caregivers are freely and lovingly given to the infant, an area defined by both Kohut and the Self Psychologists as the functioning of the Selfobject.

For Saussure the basic unit of speech is the sentence; for Bakhtin it is the "utterance". As units of language verbal sentences, utterances, belong to a responsive, interactive unit, anticipating response. The utterance marks the boundary between persons and must be directed into an already linguistically shaped context. Formulating our utterances we take account of the voices of others; we are engaged in two processes, controlling what we can say and anticipating how others may respond, over which we have no absolute control. We live in a way that is responsive both to our own position and to the position of those who are "other than" ourselves in a semiotically created "world" in which we are "placed", owning no internal sovereign territory. Our private lives are neither as private, as inner, as isolated and self-contained as we tend to assume. The "within" and the "between" are in many ways equivalents, negotiated in back and forth processes. The Russian Volosinov wrote, "By its very existential nature, the subjective psyche is to be localised somewhere between the organism and the outside world, on the borderline separating these two spheres of reality." Vigotsky states that whatever happens on the intrapsychic level has already happened on the inter- psychological, the social level, the sociogenesis of higher forms of behaviour. As John Shotter has written (*Cultural Politics of Everyday Life*, Open University Press, Buckingham 1993), every utterance has a subtext; it is an attempt to develop a sensed "thought-seed" into an utterance-flower. We are always engaged in inner speech and there are those with whom, about whom and to whom we speak in our inner speeches.

Thus my thesis is of the movement from one body to two and three-body psychology (Rickman), from Freud's neurological intrapsychic one body mental apparatus to the position of Bakhtin, who writes, "*I am conscious of myself and become myself only by revealing myself to another, through another and with the help of another ... Every internal experience ends up on the boundary. The very being of man (both internal and external) is a profound communication. To be means to communicate ... To be means to be for the other; and through him for oneself. Man has no internal sovereign territory: he is all and always on the boundary*".

REFERENCES

DAMASIO, A. R. (1995). *Descartes' Error. Emotion, Reason and the Human Brain*, London: Picador.
FUCHS, S. H. (1936). "Zum Stand der heutigen Biologie. Dargestellt an Kurt Goldstein 'Der Aufbau des Organismus'," *Imago*, 22, 210.
FREEMAN, W. J. (1995). *Societies of Brains. A study in the neuroscience of love and hate*, Hillsdale, NJ: Erlbaum.
GOLDSTEIN, K. (1995). *The Organism*, Foreword by Oliver Sacks, New York: Zone Books.
HOLQUIST, M. (1990). *Dialogism. Bakhtin and his World*, London: Routledge.
MODELL, A. "Neural Darwinism and a conceptual crisis in psychoanalysis in selectionism and the brain," *International Review of Neurology*, vol 37.
MODELL, A. (1993). *The Private Self*. Cambridge: Harvard University Press.
RIESE, W. (1958). "Freudian Concepts of Brain Functions and Brain Disease," *Journal of Nervous Mental Diseases*, 287–307.
SACKS, O. W.. "A new vision of the mind," *International Review of Neurobiology*, vol 37.
SHASKAN, D. A. and ROLTER, W. L. (1985). *Paul Schilder. Mind Explorer*, New York: Human Sciences Press.
SHOTTER, J. (1990). *Cultural Politics of Everyday Life*, Buckingham: Open University Press.
SMITH, R. (1992). *Inhibition. History and Meaning in the Sciences of Mind and Brain*, London: Free Association Books.

Erklärung der Abbildungen.

Tafel IV.

Fig. 1. Ventraler Ast einer hinteren Wurzel und das ihn begleitende Gefäss. Eine der gefässbegleitenden Fasern *gbf* ist in den ventralen Ast der hinteren Wurzel zu verfolgen. Goldpräparat. Vergrösserung 225.

spz = äusserste Spinalganglienzelle.
v A = ventraler Ast.
sz = sympathische Zelle.
ez = in den ventralen Ast eingelagerte kleine Zelle.
a A = von dem ventralen Ast abgehende Ästchen.
gbf = gefässbegleitende Faser.
z = Zweig der gefässbegleitenden Faser.
ff = feine varicöse Fasern, in die sich die gefässbegleitende Faser auflöst.

Fig. 2. Spinalganglion, ventraler Ast der hinteren Wurzel und das begleitende Gefäss. Der ventrale Ast vor der Einlagerung der ventralen Zellen gerissen. (*) Bei *C* eine Commissur zwischen zwei Nervenzellen. Goldpräparat. Vergrösserung 225.

Sp G = Spinalgefäss.
d = dorsaler,
v = ventraler Ast.
s = sympathischer Ast.
ang = angelehnte Faser.
dz' = durchziehende Faser in den sympathischen Ast.
Th = Theilung einer den dorsalen Ast kreuzenden Faser.
a A = Äste, die vom ventralen Ast abgehen.
sz = sympathische Zelle.
sdz = sympathische Doppelzelle.
G A = Gefässäste.
ff = feine Fasern, in die sich eine gefässbegleitende Faser auflöst.
Gf = Gefäss.
gbf = gefässbegleitende Faser.
zs = Spinalganglienzelle, die ihren Fortsatz in den sympathischen Ast schickt.

Fig. 3. Feines Netz varicöser Fasern auf der Pia mater. Goldpräparat. Vergrösserung 185.

PM = Pia mater.
Rmk = Rückenmark.
Wz = Wurzel.
G = Gefäss in der Pia mater.
nf = Nervenfasern, die sich in das Netz varicöser Fasern auflösen
A = ein Punkt, von welchem die sich auflösenden Nervenfasern und varicöse Fasern ausstrahlen.

Cornelius Borck

Visualizing Nerve Cells and Psychical Mechanisms
The Rhetoric of Freud's Illustrations*

"Psycho-analysis is related to psychiatry approximately as histology is to anatomy: the one studies the external forms of the organs, the other studies their construction out of tissues and cells. It is not easy to imagine a contradiction between these two species of study, of which one is a continuation of the other. To-day, as you know, anatomy is regarded by us as the foundation of scientific medicine. But there was a time when it was as much forbidden to dissect the human cadaver in order to discover the internal structure of the body as it now seems to be to practise psycho-analysis in order to learn about the internal mechanism of the mind."[1]

It is no coincidence that in 1916 Freud used the comparison between histology and anatomy to illustrate the situation of psychoanalysis and its relationship to the scientifically established related discipline of psychiatry. After all, his own scientific career had begun with studies of the histological anatomy of the nervous system. At the same time the comparison with the decidedly scientific fundamental disciplines of medicine rather than with other clinical specializations implies a claim to the scientific character of psychoanalysis. What is surprising, though, is how in making this comparison Freud gradually shifts the *tertium comparationis*. Initially he draws an analogy between the relationships of histology to anatomy on the one hand and psychoanalysis to psychiatry on the other. The comparison with anatomy which was once forbidden but has nevertheless become the basis of scientific medicine holds out the promise that psychoanalysis which at the time was still "much abused" would "in the not too distant future" become the basis of a "scientifically based psychiatry" (ibid.). However, in developing this comparison further Freud draws quite a different parallel and refers to the similarity between anatomy and psychoanalysis: just as the one "dissect[s] the human cadaver in order to discover the internal structure of the body" the other one

---

* For helpful comments I am grateful to Gerhard Fichtner, Lydia Marinelli, Andreas Mayer and John Forrester.

[1] Freud, S. (1963). Introductory Lectures on Psycho-Analysis. In: *Standard Edition*, vol. XVI, London: Hogarth Press, pp. 254–255.

explores "the internal mechanism of the mind". In this way Freud substitutes the second comparison for the first one, and only the factual parallel in the second comparison, i.e. the similarity between anatomy and psychoanalysis, makes the hope for the future plausible. Beneath the surface of the comparison of more or less established scientific disciplines is a second level of the comparison of the hidden relationship between anatomy and psychoanalysis. Both psychoanalysis and anatomy are concerned with revealing internal structures, and anatomy remains a point of reference in the discourse of psychoanalysis.

In the following I will trace the consequences of Freud's early intense concern with neuroanatomy by means of a comparative investigation of the illustrations which Freud included in his studies, from his richly illustrated works on neuroanatomy to the schematic sketches and drawings in his psychoanalytical writings.[2] The selection and formulation of my topic thus connects Freud's early papers on neuroanatomy and neurology to his subsequent and late psychoanalytic writings. In the subsequent diachronic analysis this connection will become explicit as an internal and conceptual link. The specific intertwining of psychoanalysis and anatomy which was expressed in my introductory quote from Freud's lectures on psychoanalysis runs through my discussion as well, since visualization is the goal *par excellence* of any anatomical study. It is no coincidence that Freud uses graphic representations when discussing questions of the topography of the psychical apparatus and its anatomical localization. In particular, the development of topography in psychoanalysis is reflected in a sequence of illustrations and the discussion of forms of representation. A comparison of the illustrations in Freud's writings draws our attention to the explicit and hidden connections of the anatomy of the brain, above all in his metapsychological writings.

---

[2] If one considers only those illustrations that were drawn by Freud himself, leaving aside all those which he added or copied from other sources, we are left with 28 illustrations (reprints of one and the same drawing in several publications are counted as one illustration only). 20 of these are found in his neuroanatomical and 8 in the analytical studies (2 in 1877a, 16 in 1878a, 5 in 1886c and 2 in 1891b compared to 1 in 1898b, 3 in 1900a, and 1 each in 1917c, 1921c, 1923b and 1933a). Thus there are comparatively few works on neuroanatomy with a multitude of sometimes very elaborately drawn and reproduced pictures as opposed to a great number of analytical papers which are, however, at most sparsely illustrated. Nevertheless there is a large number of works – namely the clinical studies – from Freud's neurological period in which illustrations are entirely lacking and which come, as it were, between these two latter periods. These include shorter communications on the topographical diagnostics of clinical symptoms as well as the major monographs of several hundred pages each about cerebral palsies in children. These comprehensive studies are not illustrated, even though they often discuss pathological anatomy in great detail. On the connection between these clinical studies and the formation of the psychoanalytical theory cf. Heinz G. Schott (1981). "'Traumdeutung' und 'Infantile Cerebrallähmung', Überlegungen zu Freuds Theoriebildung," *Psyche,* 35, 97–110.

Images are of central importance for Freud on many levels – from the visual language of dreams to the psychoanalytical interpretation of works of art. In my attempt to explore the inner links between neuroanatomy and psychoanalysis I will, however, limit myself entirely to those illustrations developed by Freud himself. For this reason I leave aside any discussion of illustrations done by others, even if this excludes such important studies as e.g. *Leonardo da Vinci and a Memory of his Childhood* (1910) and the paintings by the 'Wolf Man'. Moreover I will base my study only on the published illustrations without referring to any drawings in manuscripts or letters. I do make two exceptions to these principles, for in analyzing the study on aphasia the illustrations which Freud took from the writings of the neurologists discussed by him have to be taken into account, and I cannot leave unmentioned the drawings of the *Entwurf einer Psychologie [Project for a Scientific Psychology]* of 1895, which marks such an important juncture for my topic. My focus will be on revealing the rhetoric which Freud develops in dealing with his illustrations.[3] I will therefore not discuss the conceptual locus of visualizations in the theory of psychoanalysis but will concentrate instead on the language of his illustrations, following the chronological order of Freud's publications.

1. Freud's training in histological techniques and his first illustrations of neuroanatomy

In his *Selbstdarstellung [An Autobiographical Study]* Freud reports that from 1876 to 1882, hence still during his medical studies, he worked in Ernst von Brücke's physiological laboratory, where his interest in neuroanatomy was kindled: "*Brücke gave me a problem to work out in the histology of the nervous system; I succeeded in solving it to his satisfaction and in carrying the work further on my own account.*"[4] This problem resulted in Freud's first two publications on neuroanatomy (1877*a*, 1878*a*) in which he investigated the posterior nerve-roots and their relationship to the spinal ganglia and spinal cord in a primitive species of fish (Petromyzon). At first sight it may seem surprising that these decidedly histological studies originated in a physiological Institute. However, Brücke's broad research program was based on a combination of physiology and anatomy. In fact, it is precisely in anatomical studies that the foundations of functional explanations were to be looked for. In the con-

---

[3] Freud's rhetoric has repeatedly been analyzed, although without paying any closer attention to his use of illustrations; in this context we should mention Kenneth Burke (1974). *The Philosophy of Literary Form; Studies in Symbolic Action*, Berkeley; Walter Schönau (1968). *Sigmund Freuds Prosa. Literarische Elemente seines Stils*, Stuttgart; Patrick J. Mahony (1987). *Freud as a Writer*, New Haven, and Gordon Patterson (1990). "Freud's rhetoric: persuasion and history in the 1909 Clark Lectures," *Metaphor and Symbolic Activity*, 5, 215–233.

[4] Freud, S. (1959). In: *Standard Edition*, vol. XX, London: Hogarth Press, p. 10.

cept of the strictly physicalist school of Helmholtz , to which Brücke had belonged ever since his days in Berlin, the study of nerve cells under the microscope even permitted the development of the principles of a scientific psychology.[5] In addition physiological experiments were also carried out in Brücke's laboratory, e.g. by Freud's colleagues and friends Sigmund Exner and Ernst von Fleischl-Marxow.[6] For this reason it is remarkable that Freud concentrated on anatomical studies during his entire stay at Brücke's laboratory.[7]

Freud's intense concern with the technical aspects of histology can be seen in the fact that he experimented systematically with different staining techniques, devoting several publications to this topic (1879*a*, 1884*b–d*, 1887*g*). He felt he was particularly successful in developing a method of staining nervous tissue with gold chloride in such a way that the cellular organization of the tissue could be studied by means of the nerve tracts. Freud considered this work so important that he published it in three different places, among them in an English version in the journal *Brain* (1884*c*). This gold chloride staining was also an important methodological prerequisite for his studies of the anatomy of the medulla oblongata, the transition from the spinal cord to the brain (1885*d*, 1886*b–c*). Furthermore as Brücke's assistant, Freud was obligated to prepare the anatomical specimens and histological object carriers for the lectures at the university. Freud's scientific activity during those years consisted of working on visualizations; and this was true even after he left Brücke's laboratory. When working with Meynert, Freud continued his microscopic studies, publishing e.g. his studies on the medulla oblongata: "*In a certain sense I nevertheless remained faithful to the line of work which I had originally started. The subject which Brücke had proposed for my investigations had been the spinal cord of one of the lowest of the fishes (Ammocoetes Petromyzon); and I now passed on to the*

---

[5] The larger context of Freud's scientific training has been reconstructed and described especially by Siegfried Bernfeld; cf. his essays: "Freud's Earliest Theories and the School of Helmholtz," *The Psychoanalytic Quarterly* 13 (1944), 348 and "Freud's scientific beginnings," *American Imago* (Detroit) 6 (1949) 163–196. Ernest Jones also used this source in the respective sections of his Freud biography (E. Jones (1953). *The Life and Work of Sigmund Freud, vol. 1: The Formative Years and Great Discoveries, 1856–1900*, New York). See also Peter Amacher (1965), Ernst Brücke and Reflex Function, in: idem, "Freud's neurological education and its influence on psychoanalytic theory," *Psychological Issues*, 1965 4(4) (Monograph 16), pp. 9–21; and Ernst Freud, Lucie Freud and Ilse Grubrich-Simitis (Eds.) (1976). *Sigmund Freud. Sein Leben in Bildern und Texten*, Frankfurt.

[6] Cf. e.g. Exner, S. (1873/1874). "Experimentelle Untersuchungen der einfachsten psychischen Prozesse," *Archiv für Physiologie (Bonn)*, 7, 601–660 and 8, 526–537.

[7] Bernfeld reports that at Salomon Stricker's laboratory of experimental pathology Freud **did** carry out physiological experiments, albeit only with mediocre success. He therefore surmises that these negative experiences might have induced Freud to concentrate entirely upon microscopic anatomy (cf. Bernfeld, in *American Imago* 6 (1949) p. 186 [see note 5 above]).

human central nervous system."8 With the exception of his studies on cocaine, all of Freud's scientific publications from his first ten years of research are directly related to histology and techniques of visualization.

An example of the illustrations in the publications of that time for which Freud did the drawings himself is provided by the first plate from the publication *Über Spinalganglien und Rückenmark des Petromyzon* [*On the spinal ganglia and spinal cord of Petromyzon*] of 1878 (1878a) (cf. Figure 1).

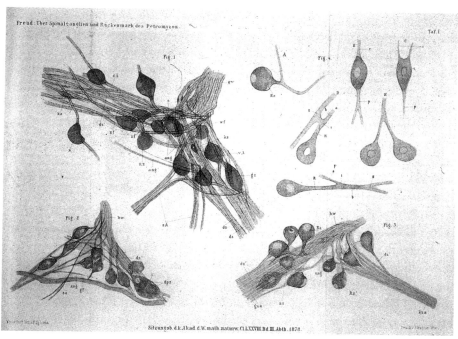

Figure 1: Table 1 from Freud, S. (1879). Über Spinalganglien und Rückenmark des Petromyzon [On the spinal ganglia and spinal cord of Petromyzon].
In: *Sitzungsberichte der kaiserlichen Akademie der Wissenschaften. Mathematisch-naturwissenschaftliche Classe*, vol. LXXVIII, Nos. 1–5, between pp. 160 and 161.

The task set by Brücke was to determine whether and how the nerve fibers entering the spinal cord through the spinal ganglion come into contact with the cells in the ganglion. In his studies Freud had observed that some of the fibers are actually continuations of the cells in the spinal ganglion while others pass through it without being involved. Moreover he had identified these nerve cells in the spinal ganglion of a ver-

---

[8] Freud, S. (1959). An Autobiographical Study. In: *Standard Edition*, vol. XX, London: Hogarth Press, p. 10.

tebrate as bipolar, which was contrary to the prevailing opinion.[9] What is of interest here, however, is less Freud's concern as a young student with the comparative neuroanatomy of his time than the visual representation of his findings. The published illustrations are drawn extremely carefully, full of precise details which show all the features of the nerve tissue investigated. At the same time they are systematically reduced to the essential features; they show only nerve cells and fibers, but not the surrounding connective tissue. Despite their very different degrees of frequency, the different forms of nerve cells observed are presented individually, completely separated from their context in the specimen, in item 4 of this plate. Nerve cells with their characteristic shape of roundish cell body and long and thin processes are found in all the plates of this study. They were the focus of Freud's attention and they are here drawn very naturalistically with the aim of rendering precisely what he had observed. The descriptive objectiveness of these illustrations is equally the product of precise observation and systematic manipulation[10]. They are therefore not comparable to photographs, even though the systematic concern with the material here remains on the level of a realistic, imitating representation.

A considerably more abstract form of systematization is found in Freud's paper *Über den Ursprung des Nervus acusticus* [*On the origin of the auditory nerve*] (1886c). Here Freud developed a form of schematic representation that was to remain characteristic of the illustration of theoretical relationships. Thematically the topic is linked to Freud's studies of the Petromyzon. The issue was once again morphological structures, the tract of nerve fibers and their connection with each other and with cellular centers. From which nuclei in the medulla or the brain do the fibers of the auditory nerve originate? The auditory nerve consists of several portions which enter the spinal cord one below another. Freud first presents illustrations of histological sections on the various layers of these portions of the auditory nerve. The gold staining technique made it possible to follow the tracts of the fibers on these levels even in the medulla, and to identify their centers of origin. However, the application of this method was not unproblematic. The advantage of staining the nerve fibers turned into a disadvantage when it was applied to nerve tissue from the central nervous system of higher mammals, because so many fibers were stained that an indistinguishable jumble resulted. Thus this visualization technique at first pro-

---

[9] Cf. Brun, R. (1936). Sigmund Freud's Leistungen auf dem Gebiete der organischen Neurologie. In: *Schweizerisches Archiv für Neurologie und Psychiatrie*, vol. 37, pp. 200–207, esp. p. 201; Bernfeld, S., op.cit., p. 129f; and Triarhou, Lazaros C.; Cerro, Manuel del (1985). Freud's contribution to neuroanatomy. In: *Archives of Neurology*, vol. 42, pp. 282–287.

[10] This statement is based on the distinction between different forms of visual objectivity as they have been identified by Lorraine Daston and Peter Galison in their study "The image of objectivity" (*Representations* 1992, 40, 81–128).

duced but another form of obscurity. However, Freud had observed that the staining method could be profitably used when studying fetal nerve tissue. Although such tissue contains only very few nerve fibers these are already characteristic of the fiber connections in the adult. The method of staining thus provided Freud with a variant of an ontogenetic principle in the field of neuroanatomy. As later in his psychological theory, the ontogenetic perspective yielded clarity in the midst of confusing complexity. The following illustration shows one of the sections prepared in this way *(Figure 2)*.

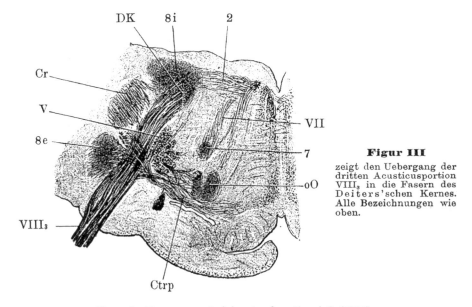

Figure 2: Neuroanatomical drawing from Freud, S. (1886).
Über den Ursprung des Nervus acusticus [On the origin of the auditory nerve].
In: *Monatsschrift für Ohrenheilkunde sowie für Kehlkopf-, Nasen-, Rachen-Krankheiten.*
N.F., vol. XX, No. 8, p. 250.

What strikes us once again are the precise details of the illustration. Even without exact knowledge of neuroanatomy we can see how the auditory nerve enters the portion of the oblongata shown here from the lower left as a thick bundle, then branches with a main bundle that leads to the core at the very top, and a thin strand running to the middle of the tissue. Freud shows precisely how certain fiber tracts cross above or below others. The various centers are more or less distinctly visible in the specimen. Although fetal tissue has been studied here, the wealth of fibers shown can still hardly be unravelled by the untrained observer. As we have mentioned this representation shows only one of the three portions of the auditory nerve. All of them together thus result in an extremely complex anatomy of the nerve. All the more surprising therefore is the clarity of the schematic illustration at the end of the paper with which Freud summarizes the different tracts of the various portions of the nerve (cf. *Figure 3*).

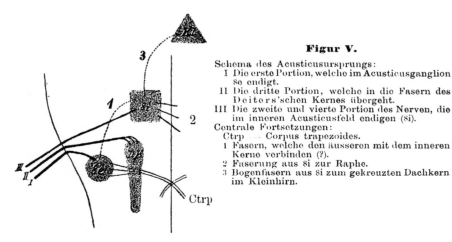

Figure 3: Schematic drawing from Freud, S. (1886).
Über den Ursprung des Nervus acusticus [On the origin of the auditory nerve].
In: *Monatsschrift für Ohrenheilkunde sowie für Kehlkopf-, Nasen-, Rachen-Krankheiten*.
N.F., vol. XX, No. 9, p. 281.

In a number of ways this drawing abstracts from the preceding illustrations: four sections are transformed into a unified schema. As a consequence the level of observation has been tilted and instead of horizontal sections a longitudinal section through the oblongata is depicted. The different graphic elements by which the nerves, nuclei and fibers are represented still resemble the differences of their physical substrata in the specimen, but at the same time they are abstract formal elements that enhance the descriptiveness of the illustration. Everything has been reduced to the essential, the anatomy has turned into something akin to part of a circuit. The schematic representation of the origin of the auditory nerve underscores in a special way the functional aspects of the anatomy. At the time such schematic representations were quite common in works on neuroanatomy as will become clear from the discussion of the illustrations in the study on aphasia below. The paper on the origin of the auditory nerve shows especially clearly how a schematic representation evolves out of a series of anatomical depictions. The schematic representation continues to refer closely to the anatomical visualization, but it displays a structure that transcends the individual levels of the sections of the specimens.

2. The Illustration of 'Non-Depictability' in Freud's Study on Aphasia

On the Interpretation of the Aphasias, a critical study is Freud's critical investigation of the new concept of the localization of different aspects of speech in different specific areas of the cerebrum. As is well known, Freud was opposed to an absolute interpretation of the localization principle and the implied anatomical dissociation of the speech apparatus. In contrast to this he formulated a functional approach which essentially considered speech an integrated product of the nervous system that cannot be broken down in such a way that anatomical loci and partial speech competencies could be matched. Freud dealt with the contemporary neurological ideas and particularly with Carl Wernicke on the basis of an intensive study of the literature and his general neurological experience but without adding any clinical findings of his own. This fact explains why this work contains numerous illustrations from the writings discussed by Freud but no *anatomical* illustrations drawn by him. Altogether only two of the ten figures in this book were designed by Freud, one of them illustrating Freud's speech model and the other one its incompatibility with anatomical localizations.

The illustrations by other authors included in the study on aphasia may be more or less divided into the two groups of "anatomical depictions" and "functional diagrams", which we encountered already in Freud's illustrations in his work about the auditory nerve. Wernicke's representation from *Der aphasische Symptomencom-*

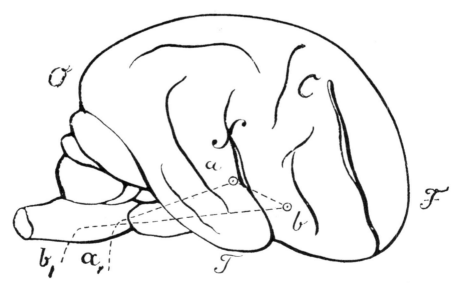

Figure 4: Wernicke's anatomical illustration of aphasia as reprinted by Freud in: Freud, S. (1891). *Zur Auffassung der Aphasien [On Aphasia]*, Leipzig, Wien: Franz Deuticke, p. 5.

plex with its distinction between a motor and a sensory speech center (cf. *Figure 4*) was a sort of icon of the discourse on the localization of the parts of the speech apparatus. A few rough lines sketch a sort of side view of the brain, in which lower case letters indicate the speech processes: along the dashed line α an acoustic impression reaches the auditory speech center in the brain (a), passes from there to the motor speech center (b) which finally results in a speech act along the pathway β. Although this representation with the excitation path drawn in goes beyond mere anatomy, anatomical questions clearly predominate, and we therefore include it here as an example of an anatomical representation. The functional diagrams, on the other hand, operate with an entirely different register of visual means of expression as the following illustration shows (cf. *Figure 5*).

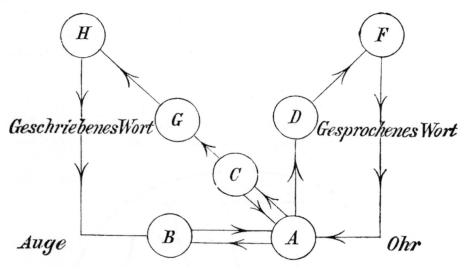

Figure 5: Grashey's functional diagram of aphasia as reprinted by Freud in: Freud, S. (1891). *Zur Auffassung der Aphasien [On Aphasia]*, Leipzig, Wien: Franz Deuticke, p. 36.

This schema from a work by Grashey considers not only the facts illustrated by Wernicke, but also the fact that speech is coupled with both acoustic and optical impressions (e.g. in reading the latter may also lead to speech). Instead of inscribing such additional links as hypothesized nerve fibers leading to the visual center of the brain, Grashey developed a purely graphical and abstract representation that foregoes any depictive similarity to a brain. The presumed centers are here schematically reduced to circles and the nerves connecting them to arrows, whose directions indicate the course of excitation during the processing of speech. The junctures arising in this way are an amalgamation of functional units and anatomical loci; they refer to both clinically distinguishable functional sections and neuroanatomical centers. Some of these junctures referred to as "centers" had clearly defined anatomical correspondences, others were postulated centers of a still unknown localization. Although they did not even remotely resemble a brain and despite the complexity of the functional relationships represented, these diagrams nonetheless referred precisely and implicitly to neuroanatomy. The study on aphasia also documents that such graphically reduced representations consisting of circles, lines and arrows were quite common in neurology at the time. When Freud used them later to represent psychical mechanisms he continued this tradition.

The two illustrations drawn by Freud himself for this study are labeled "Psychological schema" and "Anatomical schema". He thus takes the two groups of representations as his starting point, placing his own illustrations into the interpretative contexts of these visualizing techniques. However, in both cases he undercut the appearance he evoked, countering it with a critical intent. At first sight the "psychological schema" (cf. *Figure 6*) suggests a schema like Grashey's.

With similar graphical means, circles and lines, a web of the interrelatedness of the parts of speech competency has been sketched.[11] However, Freud here illus-

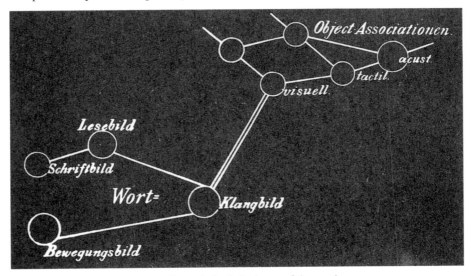

Figure 6: Freud's psychological schema of the word concept
in: Freud, S. (1891). *Zur Auffassung der Aphasien [On Aphasia]*, Leipzig, Wien: Franz Deuticke, p. 79.

---

[11] The inverted color scheme – white on black – in the German edition (in contrast to the English edition) sets these two illustrations off from the text, but this also applies to Lichtheim's illustrations reprinted by Freud (cf. 1891*b*, pp. 7 and 8).

trates a psychological concept of speech which is based on an entirely different notion of an interlinking of ideas of word-presentations and object associations.[12] The point of the diagram is precisely that it *cannot* be mapped onto a brain surface. What is shown are functionally differentiated (partial) aspects of the totality of the speech apparatus, which resulted for Freud from a study of the psychology of speech without any concern for the possibility of their precise anatomical attribution. Even Freud's basic difference between the concept of word-presentation and object-association, by which he understands language essentially as a system of symbols in which the signified and the signifier have to be constantly relinked with each other, was outside the localization discourse that went on at the time. It is also remarkable how the openness of object associations presumed by Freud is indicated in the illustration by an empty, unlabeled circle and links towards the outside. These details mark the conceptual precision with which Freud had designed this scheme. The 'Psychological schema' thus illustrates the complexity of the human ability of speech in contrast to the simplified assumptions of the authors discussed by Freud.

It is only consistent that when Freud added an 'Anatomical schema' to the text a few pages later he described its purpose in the caption as a didactic one, as "demonstrating how the appearance of speech centers is created" (cf. *Figure 7*). Freud used this illustration for a critical deconstruction of the assumption of anatomically localizable speech centers. Instead of one or several speech centers Freud believed that there was a whole field of speech associations which in its entirety was responsible for speech and could not be broken down further anatomically. Freud pursued his critical goal with remarkable means: although it is supposed to be an anatomical schema in which, as the caption says, "the auditory and visual receptive fields and the motor areas for the muscles serving articulation and writing are represented by circles", it purposefully avoids any similarity to the representation of a brain. The caption further talks of the area in which the association tracts "are crossed by the corresponding fascicles from the other hemisphere" which "becomes a centre for the respective associative element". The areas shown in the figure where two radiating fascicles cross are, however, intentionally left vague. They are crossings of the fascicles of a single area of the cortex with fascicles from an indeterminate

---

[12] Later Freud took this theory of speech as a starting point, especially in his article on the 'Subconscious'. This context which reveals very interesting links between neurological and psychoanalytical concepts in Freud which I cannot discuss further here was first pointed out by E. Stengel (1954/55). "Die Bedeutung von Freuds Aphasiestudie für die Psychoanalyse", *Psyche*, 8, 17–24. Cf. also Sebastian Geppert (1974). "Die Funktion der Sprache in Freuds 'Zur Auffassung der Aphasien'", *Jahrbuch der Psychoanalyse*, Beiheft 2, 7–38; and John Forrester (1980). *Language and the Origins of Psychoanalysis*, Cambridge.

Visualizing Nerve Cells and Psychical Mechanisms | 69

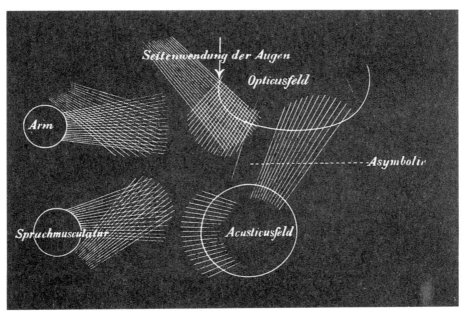

Figure 7: Freud's anatomical schema of the area of the speech associations in: Freud, S. (1891). *Zur Auffassung der Aphasien [On Aphasia]*, Leipzig, Wien: Franz Deuticke, p. 83.

background. In the center of the drawing there would in fact be an area in which all these radiating fascicles would cross and which according to Freud is the field of speech associations. This field of speech associations which in Freud's opinion cannot be clarified any further anatomically has been left blank in the centre of the figure.

This schema thus follows a logic of showing by receding and leaving things open. This complicated logic of representation was further compounded by Freud's addition of details that hardly belong here logically. E.g. at the top of the schema near the visual receptive field a semicircular line with an arrow and the words "Seitenwendung der Augen" ["deviation of the eyes"] has been added which introduces a dynamic element into the drawing whose connection to the rest of the schema remains unclear. On the right side of the drawing a dashed line designated as "Asymbolie" ["asymbolia"] indicates a half-hearted attempt to mark a certain pathological disturbance in a quasi-anatomical manner. The explanatory legend expressly states that "The crossed connections of the auditory receptive field have been omitted from the schema to avoid confusion, and also because of the uncertainty of the connections between the auditory receptive field and the sensory speech centre." At this point the drawing suddenly seems to strive for anatomical precision.[13] The one-sided auditory receptive field of this illustration has later reverberations in the one-sided "cap of hearing", which the figure of the mental apparatus will wear in *The Ego and the Id* (see Figure 16 below).

This brief discussion should suffice to illustrate how Freud carried out his criticism of the localization principle especially also on the level of visual representation. The 'Anatomical schema' does not provide an anatomical representation of aphasia. Instead, Freud created this illustration precisely in order to illustrate the impossibility of depicting the psychological complexity of speech by means of anatomical localization. The failure of the illustration as an anatomical representation is part of its logic of representation. In this way the 'Anatomical schema' is an example of a visualization strategy that can only be called rhetorical.[14] A picture illustrates the non-depictability of a relationship. Freud continued the discussion of the anatomical localization of speech by special graphical means. The 'Anatomical schema' is purposefully inadequate in order to visualize the problematic nature of a localizing topographical visualization. According to the logic of the argumentation in the study on aphasia the schema is successful precisely because of its failure, in sharp contrast to the 'Psychological schema' which visualizes the connections developed in the text by graphical means. With regard to the anatomy Freud designed an aporia of visualization in order to win the argument on the performative level.

The two illustrations created by Freud himself for his study on aphasia show a tension between functional and topographic representation which is also seen in his subsequent illustrations. In particular in discussing the structure and function of the

---

[13] The superb polemics of Freud's study on aphasia as a whole is an equally strong proof of the fact that the details observed here have been included in the figure on purpose in the sense of a rhetoric that needs to be reconstructed, just like the precision of the details of the other figure drawn by Freud himself, the 'Psychological schema'. In connection with his study on aphasia Freud wrote several shorter publications in the form of articles for encylopaedias, in the context of which he also had his 'Psychological schema' reprinted [in the *Diagnostisches Lexikon für praktische Ärzte*, vol. 1, 1893; cf. Oswald U. Kästle (1987). Einige bisher unbekannte Texte von Sigmund Freud aus den Jahren 1893/94 und ihr Stellenwert in seiner wissenschaftlichen Entwicklung. In: *Psyche*, vol. 41, pp. 508–528, fig. on p. 522; see also Johann G. Reicheneder (1994). 'Lokalisation': Ein bisher unbekannt gebliebener Beitrag Freuds zu Villarets Handwörterbuch der gesamten Medizin. In: *Jahrbuch der Psychoanalyse*, vol. 32, pp. 155–182].

[14] It is the intention of this paper to understand the purposeful design of an illustration as part of a rhetorical action. In this instance I use 'rhetorical' in a loose allusion to e.g. the 'rhetorical question', which is in fact no longer a question. The context studied here, however, is more complex, since the 'Anatomical schema' is no more an anatomical depiction than the rhetorical question is a question, but this does not exhaust the function of the rhetorical visualization: this illustration addresses precisely the problem of the 'depictability' of the facts discussed. The rhetorical visualizing is thus not only a stylistic device but an objectively motivated strategy that opens up a further level for the discussion of the issue.

psychical apparatus Freud repeatedly addressed the problem of representation and developed a series of original solutions precisely in the same field of tension between functional diagram and topographic representation that is obvious here. The question of the localization and structure of the psychical apparatus did not become obsolete by the graphical-rhetorical aporia in Freud's localization of linguistic abilities.

3. Nerve cells as a functional relay of psychical mechanisms

Before looking in detail at the progression sketched above from Freud's study on aphasia to the topography of the psychical apparatus another line of transformation of neuroanatomical illustrations must be pointed out which can be followed exemplarily on the basis of the sketches included in Freud's manuscript of the *Entwurf einer wissenschaftlichen Psychologie* [*Project for a Scientific Psychology*] of 1895, which anticipates the basic figure of the graphic representation of psychical mechanisms.[15] On the basis of two sketches from the *Project* we can see paradigmatically how morphological-anatomical representations were replaced by a psychological functional schema which freed itself more and more from the morphological depictions.

The first sketch shows nerve cells with their projections that are reminiscent of Freud's very first illustrations in his study of the Petromyzon (cf. *Figure 8*). What was important to Freud here were the ramifications, for it is they that make it possible to alter the course of excitation indicated by arrows in the reflex arc from a) to b), redirecting part of the sum of excitation (indicated as $Q\eta$ on the left) to the neutrons α–δ. On the basis of the neurophysiological principle of facilitation the sketch illustrates the concept of the ego inhibition. To put it briefly, Freud here develops a concept in which the diversity of the performance of the nervous system and especially the psychical phenomena are derived from the ramifications of intercon-

---

[15] Several commentators view the *Entwurf* as evidence that psychoanalysis originated in 19th century psychophysical neurophysiology, cf. Amacher, Peter (1965). "Freud's neurological education and its influence on psychoanalytic theory", *Psychological Issues*, 4 (4), Monograph 16; Fancher, Raymond E. (1971). "The neurological origin of Freud's dream theory", *Journal of the History of the Behavioral Sciences*, 7, 59–74; Sulloway, Frank J. (1979). *Freud, Biologist of the Mind*, New York; Heynick, Frank (1985). "Dream dialogue and retrogression: neurolinguistic origins of Freud's 'replay hypothesis'", *Journal of the History of the Behavioral Sciences*, 21, 321–341. In fact a comparison of the *Entwurf* e.g. with the *Entwurf zu einer physiologischen Erklärung der psychischen Erscheinungen* (Vienna 1894) published a year earlier by Freud's colleague Sigmund Exner is striking and points to lines of tradition along which Freud developed some of the concepts of his *Entwurf*. To me, however, Freud's critical concern with the neurology of his time as it is clearly seen in his study on aphasia seems to be more decisive for his psychoanalytical writings than the ultimately failed attempt of the *Project* to amalgamate psychology and neurophysiology (a similar argument is presented by Mark Solms and Michael Saling (1986). "On psychoanalysis and neuroscience: Freud's attitude to the localizationist tradition", *International Journal of Psycho-Analysis*, 67, 397–416). The fact that Freud never revised this text – and its sketches – for a publication calls for caution in dealing with this material which must not be measured by the same standard as the carefully elaborated illustrations of his study on aphasia.

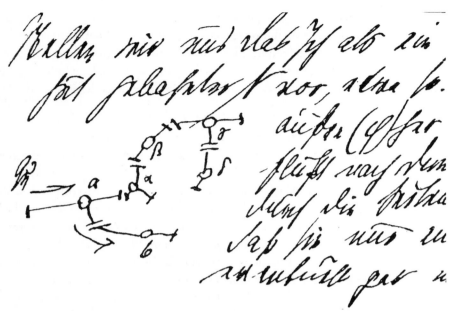

Figure 8: Part of a page from the manuscript of Entwurf einer Psychologie
[Project for a Scientific Psychology]
In: Freud, S. (1987). *Gesammelte Werke: Nachtragsband, Texte aus den Jahren 1885–1938.*
Frankfurt am Main: S. Fischer Verlag, p. 419.

nected nerve cells. Very much in the sense of Brücke's research program, the investigation of the morphology of the nervous system reveals information about the functions enabled by it. Even though Freud illustrates here quite generally the principle of excitation being transmitted from one nerve cell to another rather than drawing any specific anatomical specimen, the illustration stays on the level of the morphology of individual nerve cells.

The second sketch uses similar graphical means of representation, namely circles, lines and arrows. It does not illustrate any morphology, however, but the psychical mechanism of a process of repression (cf. *Figure 9*). The circles which in the first sketch represent the bodies of the nerve cells are here used to represent ideas, some of which are conscious perceptions (the filled circles) while others are recollections that are not linked to consciousness (the unfilled circles).[16] With the

---

[16] The filled circles are designated as 'Shop-assistants, Laughing, Clothes, Sexual release' and the unfilled circles underneath as 'Shopkeeper, Clothes, Assault' and those on the right as 'Being alone, Shop, Flight'. The schema explains the at first sight unintelligible flight of one of Freud's female patients from a shop because the shop-assistant had laughed at her, as a reaction to the unconscious recollection of a sexual assault which she was now trying to escape from in the shop.

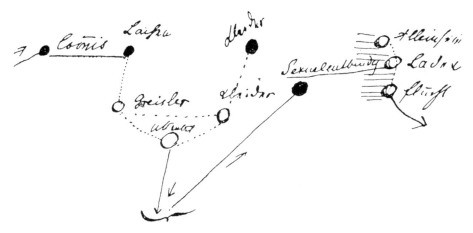

Figure 9: Sketch of psychical mechanism from the manuscript of Entwurf einer Psychologie [Project for a Scientific Psychology]
In: Freud, S. (1987). *Gesammelte Werke: Nachtragsband, Texte aus den Jahren 1885–1938*. Frankfurt am Main: S. Fischer Verlag, p. 446.

help of this functional diagram an action that is at first unintelligible can be explained as a connection of conscious and unconscious ideas. In this sketch ideas are linked with each other in analogy to the diagram of the nerve cells. By using identical graphical means the two sketches show that an association in the psychic sphere is similar to a facilitation on the level of the nerve cells. The psychical mechanism takes the place of nerve cell topography.

Freud used the same inventory of graphic patterns to describe psychical mechanisms also in several diagrams of his published writings. They employ the same uniform pictorial language of geometric figures, connecting lines and arrows that we could show on the basis of the sketches in the *Project* to relate directly to neuroanatomical ideas. The first of these illustrations, from the brief text *Zum psychischen Mechanismus der Vergeßlichkeit [The Psychical Mechanism of Forgetfulness]* of 1898 illustrates how once Freud remembered the wrong names 'Botticelli' and 'Boltraffio' instead of the name 'Signorelli', which he had forgotten (cf. *Figure 10*).

The schematic diagram uses the same topographical order from top to bottom corresponding to the conscious and unconscious as the sketch in the *Project* just discussed. It illustrates the links which Freud explains sentence by sentence in the text, thus providing a clearly arranged summary of the train of thought in a uniform diagram. This same illustration was reused by Freud for Chapter 1 of the *Psychopathologie des Alltagslebens [Psychopathology of Everyday Life]* of 1901, thus in a way framing *The Interpretation of Dreams*.

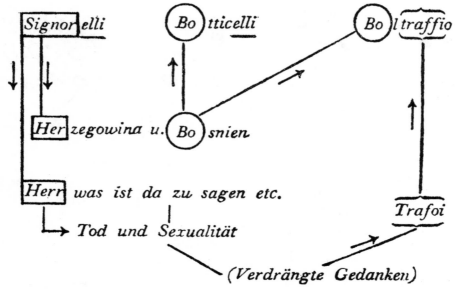

Figure 10: Illustration from Freud, S. (1972). Zum psychischen Mechanismus der Vergesslichkeit. [The Psychical Mechanism of Forgetfulness]
In: Freud, S. *Gesammelte Werke*, vol. I. Frankfurt am Main: S. Fischer Verlag, p. 524.

A second graphic representation of a similar type is found in Freud's *On Transformations of Instinct as Exemplified in Anal Erotism* of 1917 (cf. *Figure 11*). There it provides a resumé, but it remains incomprehensible by itself and is also not really conclusively explained by Freud. He does not e.g. give any reasons for different links, full lines, dotted lines and double lines, or for the open box around "Geld" ["Money"]. The designation "Narzissmus, Kastrationskomplex" ["Narcissism, Castration Complex"] at the connecting line between Anal erotism and Penis (with arrows pointing towards one another) or the arrow leading out of the drawing on the lower left designated "Trotz" ["Defiance"] are likewise not explained. Some links in the illustration can be understood on the basis of the text, but in general it appears overloaded and tends to increase the confusion that had arisen from the condensed description in the text instead of resolving it. All the more surprising are therefore Freud's comments with which he incorporates this illustration in the text. First the didactic intention of the graphic representation is expressly emphasized – and subsequently its failure freely admitted.

"*I feel sure that by this time the manifold interrelations of the series – faeces, penis, baby – have become totally unintelligible; so I will try to remedy the defect by presenting them diagramatically […]. Unfortunately, this technical device is not sufficiently pliable*

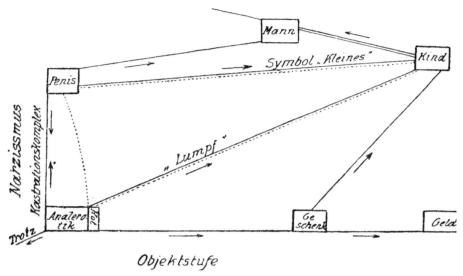

Figure 11: Illustration from Freud, S. (1969). Über Triebumsetzung, insbesondere der Analerotik. [On Transformations of Instinct as Exemplified in Anal Erotism] In: Freud, S. *Gesammelte Werke*, vol. X. Frankfurt am Main: S. Fischer Verlag, p. 408.

for our purpose, or possibly we have not yet learned to use it with effect. In any case I hope the reader will not expect too much from it."[17]

The graphic representation is meant to achieve a new degree of clarity, but it fails to reach this self-proclaimed goal. When Freud, although openly deploring its unsatisfactoriness, nevertheless included this illustration in the printed work, its use must be seen in its very insufficiency. The illustration is the extreme example of a rhetorical visualization whose meaning is found entirely in the negative: Because the text itself had become enigmatic an illustration was supposed to illuminate the train of thought. The schema shows above all that the facts *cannot* be expressed graphically.

After 1916 Freud rarely used graphic representations. Only three more illustrations were published, the two versions of the psychical apparatus from *The Ego and the Id* and the *New Introductory Lectures on Psycho-Analysis* and the diagram of the formation of the ego ideal in *Group Psychology and the Analysis of the Ego* of 1921, which we want to briefly discuss here, since it likewise illustrates a psychical mechanism (cf. *Figure 12*).

Once again the schematic representation is placed at the end of the development of a thought and at the end of a chapter. It is clearly more abstract in its formal

---

[17] Freud, S. (1955). On Transformations of Instinct as Exemplified in Anal Erotism. In: *Standard Edition*, vol. XVII, London: Hogarth Press, pp. 131–132.

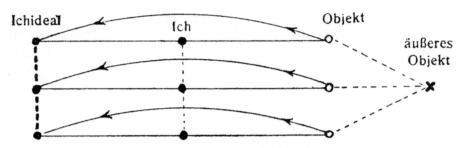

Figure 12: Illustration from Freud, S. (1972). Massenpsychologie und Ich-Analyse. [Group Psychology and the Analysis of the Ego] In: Freud, S. *Gesammelte Werke*, vol. XIII. Frankfurt am Main: S. Fischer Verlag, p. 128.

language, using very sparse graphical means. Precisely for this reason its effect is clear and convincing, although the merging of the individual ego ideals into a single external object, which is indicated by the dotted lines, can only be followed intellectually. The psychical mechanism transcends its graphic representation.

4. The representations of the psychical apparatus in the Interpretation of Dreams

The study of the graphic representations of psychical mechanisms has led us from the sketches of the *Project* into the midst of Freud's psychoanalytical writings. In this way an important group of illustrations from the *Interpretation of Dreams* has been left out, in which Freud visualizes the function of the psychical apparatus. The description of the various systems of the psychical apparatus and their topographical organization led Freud to a discussion of the anatomical localization of the psychical apparatus – a topic which figured prominently already in the study on aphasia. In the famous seventh chapter of the *Interpretation of Dreams* Freud includes three very abstract and schematic drawings to explain the psychical apparatus (cf. *Figure 13*).

With the most economical graphical means and in radical simplicity Freud sketches the fundamental schematic picture of psychical activity: Through several intermediate elements that are not defined in detail a perception (W) leads to a motor effect (M). Even if the main idea of this schema is still linked to the physiology of the reflex action which was also the basis for the *Project,* all affinities to a possible neuroanatomical substratum have now been eliminated. Even the visual language of this schematic picture does not permit any associations to nerve cells. Here a psychical mechanism is developed without any reference to anatomy. The next two illustrations adhere to the same basic scheme and only the middle elements have been altered to illustrate the new theoretical concepts introduced. The second of the three illustrations adds memory pictures between the W and the M poles,

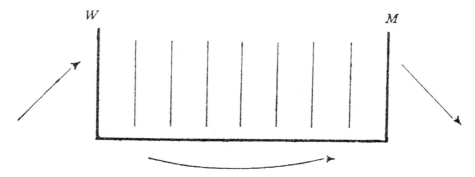

Figure 13: First diagram in Freud, S. (1973). Die Traumdeutung.
[The Interpretation of Dreams]
In: Freud, S. *Gesammelte Werke*, vols. II/III. Frankfurt am Main: S. Fischer Verlag, p. 542.

which should be carefully distinguished from the perceptions, while the last of the three schematic drawings includes a sphere of the subconscious ['Ubw'] and preconscious ['Vbw'] as well (cf. *Figures 14 and 15*).

Thus the three schematic drawings reflect the development of the theory in the text. At the same time they may also be interpreted from an ontogenetic perspective as a differential development of the psychical apparatus. In addition, the last of these schematic drawings serves to introduce visually the concept of 'regression' as a reversal of the functional (and reading) direction of the schematic picture.[18] In his *Interpretation of Dreams* Freud developed a graphic representation of the psychical apparatus that seeks to exclude any reference to anatomy. Nevertheless the question forces itself upon us as to where this apparatus and its parts should be localized. In fact Freud discussed the issue of an anatomical localization of the psychical apparatus in just this section of the *Interpretation of Dreams*, albeit only negatively by strictly rejecting any attempt at a neuroanatomical localization:

"*What is presented to us in these words is the idea of psychical locality. I shall entirely disregard the fact that the mental apparatus with which we are here concerned is also known to us in the form of an anatomical preparation, and I shall carefully avoid the temptation to determine psychical locality in any anatomical fashion. I shall remain*

---

[18] "*The only way in which we can describe what happens in hallucinatory dreams is by saying that the excitation moves in a backward direction. Instead of being transmitted towards the motor end of the apparatus it moves towards the sensory end and finally reaches the perceptual system. If we describe as "progressive" the direction taken by psychical processes arising from the unconscious during waking life, then we may speak of dreams as having a "regressive" character. This regression, then, is undoubtedly…*" Freud, S. (1958). Interpretation of Dreams. In: *Standard Edition*, vol. V, London: Hogarth Press, p. 542.

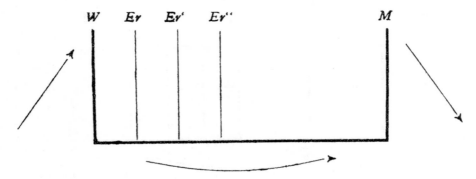

Figure 14: Second diagram in Freud, S. (1973). Die Traumdeutung.
[The Interpretation of Dreams]
In: Freud, S. *Gesammelte Werke*, vols. II/III. Frankfurt am Main: S. Fischer Verlag, p. 543.

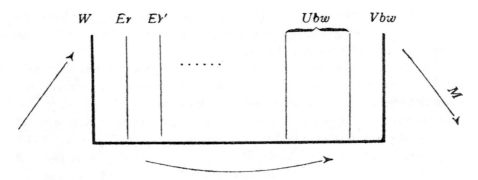

Figure 15: Third diagram in Freud, S. (1973). Die Traumdeutung.
[The Interpretation of Dreams]
In: Freud, S. *Gesammelte Werke*, vols. II/III. Frankfurt am Main: S. Fischer Verlag, p. 546.

upon psychological ground, and I propose simply to follow the suggestion that we should picture the instrument which carries out our mental functions as resembling a compound microscope or a photographic apparatus, or something of the kind. On that basis, psychical locality will correspond to a point inside the apparatus at which one of the preliminary stages of an image comes into being. In the microscope and telescope, as we know, these occur in part at ideal points, regions in which no tangible component of the apparatus is situated."[19]

---

[19] Freud, S. (1958). Interpretation of Dreams. In: *Standard Edition*, vol. V, London: Hogarth Press, p. 536.

This rhetorically extremely skillful text nevertheless reveals a hidden tension in Freud's position on the localization discourse: at the very beginning Freud concedes that he knows the psychical apparatus also as an "anatomical preparation", which of course implies much more than just an anatomical localization. Here the text is colored by the fact that Freud had long been intensely involved with anatomical specimens and neurophysiology at Brücke's und Meynert's laboratories. The issue of the localization of the psychical apparatus is to be "carefully avoided", but as an object of comparison for the psychical apparatus Freud introduces the microscope of all things which once again refers back to anatomical specimens and Freud's microscopic studies. The comparison with a microscope finally leads to the assumption of an "ideal point" which does not correspond to any material substratum. This is the place where Freud localizes his psychical apparatus. Freud's radical rejection of the localization concept of the anatomy of the brain can be pursued all the way to the metaphorics of his formulations.

At the end of the *Interpretation of Dreams* Freud discusses the issue of localization once more, resuming the same metaphorics of virtual sites, but now localizing the psychical processes in a quasi-neuroanatomical manner 'between' *them [i.e. the organic elements of the nervous system], where resistances and facilitations [Bahnungen] provide the corresponding correlates* (loc.cit. p. 611).[20] The virtual site in the 'between' of the neuroanatomical structures is clearly defined neurophysiologically with the "facilitations" and "resistances", taking up concepts from Freud's *Project*. Freud conceives of a site which cannot be localized precisely in any one area of the brain, but is nonetheless neurophysiologically defined. The radical rejection of the localization principle of brain anatomy is the surface below which Freud continued his argument with neurophysiology and the localization principle. This was to occupy Freud to the end of his life. In Freud's thinking the anatomy of the brain and psychoanalysis are in constant tension. After first addressing the issue in the *Interpretation of Dreams*, Freud kept looking for possible ways of mediating between an autonomy of the psychical and its anatomical localization until he wrote *An Outline of Psycho-Analysis* in 1938.

Freud's most decisive rejection of an anatomical localization of the psychical apparatus is found in the study about *The Unconscious* of 1915: *Research has given*

---

[20] The entire passage reads as follows: *We can avoid any possible abuse of this method of representation by recollecting that ideas, thoughts and psychical structures in general must never be regarded as localized in organic elements of the nervous system but rather, as one might say, between them, where resistances and facilitations [Bahnungen] provide the corresponding correlates. Everything that can be an object of our internal perception is virtual, like the image produced in a telescope by the passage of light-rays. But we are not justified in assuming the existence of the systems (which are not in any way psychical entities themselves and can never be accessible to our psychical perception) like the lenses of the telescope which cast the image."* Freud, S. (1958). Interpretation of Dreams. In: *Standard Edition*, vol. V, London: Hogarth Press, p. 611, the emphasis is the same as in the original.

irrefutable proof that mental activity is bound up with the function of the brain as it is with no other organ. We are taken a step further – we do not know how much – by the discovery of the unequal importance of the different parts of the brain and their special relations to particular parts of the body and to particular mental activities. But every attempt to go on from there to discover a localization of mental processes, every endeavour to think of ideas as stored up in nerve-cells and of excitations as travelling along nerve-fibres, has miscarried completely. The same fate would await any theory which attempted to recognize, let us say, the anatomical position of the system Cs. – conscious mental activity – as being in the cortex, and to localize the unconscious processes in the subcortical parts of the brain.[21] *The Unconscious* did not remain Freud's last word on topography, he later localized the psychical apparatus in the cortex.

5. The psychical apparatus as a 'cortical vesicle'

Freud's renewed concern with the anatomy of the brain began with the reworking of topography in *Beyond the Pleasure Principle* and led to the famous representation of the psychical apparatus as a vesicle (cf. *Figure 16*) in *The Ego and the Id*. At first sight the pictorial language of this representation seems quite abstract. In an approximate circle are drawn an 'Ich' ['Ego'], 'Es' ['Id'], 'Vbw' (= vorbewußt [preconscious] and 'Vdgt' (= verdrängt [repressed]) as areas loosely separated from one another, of which the Id takes up the greatest amount of space. In one pole the circle-like figure is topped by the perceptive consciousness ('W-Bw') and right next to it it wears a 'cap of hearing', as Freud has called the box designated as 'akust' in the accompanying text. 'Ego' and 'Id' are emphasized by the size and type of their lettering. The repressed is divided from the 'Ego' by a sharp gap. From afar and with a lot of imagination this figure might be reminiscent of a lateral view of the brain. Freud himself has established such a bridge to anatomy when he wrote in the text: "*We might add, perhaps, that the ego wears a 'cap of hearing' – on one side only, as we learn from cerebral anatomy. It might be said to wear it awry.*"[22] Thus at this point the fantastic anatomy of the psychical apparatus is supposed to have a certain neuroanatomical precision. With the irony of this statement Freud immediately undermines its credibility, but in any case anatomy is at least mentioned here. In addition on the following page the text refers to neuroanatomy in a way that is not at all ironical: *"If we wish to find an anatomical analogy for [the ego] we can best identify it with the 'cortical homunculus' of the anatomists, which stands on its head in the cortex, sticks up its heels, faces backwards and, as we know, has its speech-area on the left-hand*

---

[21] Freud, S. (1957). The Unconscious. In: *Standard Edition*, vol. XIV, London: Hogarth Press, p. 174.

[22] Freud, S. (1961). The Ego and the Id. In: *Standard Edition*, vol. XIX, London: Hogarth Press, p. 25.

Visualizing Nerve Cells and Psychical Mechanisms | 81

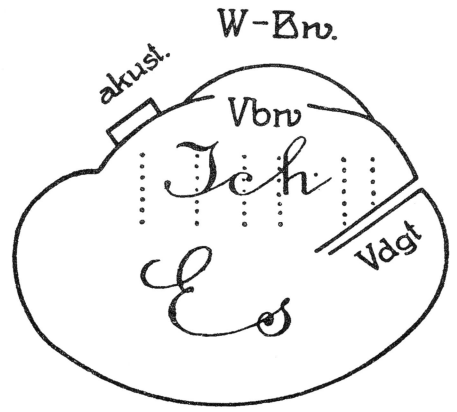

Figure 16: The psychical apparatus. In: Freud, S. (1972). Das Ich und das Es. [The Ego and the Id] In: Freud, S. *Gesammelte Werke*, vol. XIII. Frankfurt am Main: S. Fischer Verlag, p. 252.

side."[23] With the 'cortical homunculus' Freud explicitly refers to the localization discourse.

As is well known, Freud's *The Ego and the Id* serves to introduce the terminology of the 'ego', 'id' and 'super-ego' and to establish a new topographical attribution of these areas to the systems of subconscious, preconscious and perceptive consciousness in the psychical apparatus. To this extent the graphic representation illustrates the development of theory. The references to cerebral anatomy, however, give rise to cross connections to the discussion of questions of anatomy in *Beyond the Pleasure Principle*, which not only make Freud's reference to the anatomy of the

---

[23] Loc.cit. p. 26.

brain in *The Ego and the Id* seem less erratic but also offer a surprising key to the formal language of the graphic representation.

Beyond the Pleasure Principle is a further example of the fact that in Freud the development of psychoanalytical theory was paralleled by an ongoing concern with neuroscientific questions: Freud offers a detailed discussion of the physiological functioning of the nervous system in the same chapter of *Beyond the Pleasure Principle* that prepares the assumption of the death instinct. The chapter begins with the famous words: *What follows is speculation, often far-fetched speculation*", which refer to the assumption of the death instinct as well as to the specific biology of the nervous system which Freud develops here. Once more his starting point was the anatomy of the brain to which he here referred in a positive way, though:

What consciousness yields consists essentially of perceptions of excitations coming from the external world and of feelings of pleasure and unpleasure which can only arise from within the mental apparatus; it is therefore possible to assign to the system Pcpt.-Cs. *a position in space. It must lie on the borderline between outside and inside; it must be turned towards the exernal world and must envelop the other psychical systems. It will be seen that there is nothing daringly new in these assumptions; we have merely adopted the views on localization held by cerebral anatomy, which locates the 'seat' of consciousness in the cerebral cortex – the outermost, enveloping layer of the central organ.*[24]

The ego as the local link between external and internal impressions had to lie on the borderline between outside and inside, thus becoming localizable neuroanatomically in the cerebral cortex. This localization of consciousness in the cortex was, however, only the prelude to a consideration on evolutionary theory which stressed the special features of the cortex: "*Let us picture a living organism in its most simplified possible form as an undifferentiated vesicle of a substance that is susceptible to stimulation. Then the surface turned towards the external world will from its very situation be differentiated and will serve as an organ for receiving stimuli. Indeed embryology, in its capacity as a recapitulation of developmental history, actually shows us that the central nervous system originates from the ectoderm; the grey matter of the cortex remains a derivative of the primitive superficial layer of the organism ..."*[25]

The mental apparatus of the developed brain is the late derivative of the 'undifferentiated vesicle of a substance that is susceptible to stimulation'. Freud's ideas on neurophysiology and evolutionary theory in *Beyond the Pleasure Principle* reveal precisely the blending of mental apparatus and primordial vesicle which found its graphic expression in the figure of *The Ego and the Id*. It is probably also such neuroscientific

---

[24] Freud, S. (1955). Beyond the Pleasure Principle. In: *Standard Edition*, vol. XVIII, London: Hogarth Press, p. 24.
[25] Loc.cit. p. 26.

considerations as they appear in the references to neuroanatomy included in the text that motivated Freud to develop a morphological representation of the psychical apparatus in *The Ego and the Id* – in contrast to the *Interpretation of Dreams*.

The comparison of the brain with a vesicle has a prominent precursor in the literature which must have certainly been known to Freud and which implicitly links this passage to the late 19th century neuropsychiatric discourse. It is found in Theodor Meynert's – Freud's superior during the period of his clinical activity – *Psychiatrie. Klinik der Erkrankungen des Vorderhirns*: "*In the way in which the cortical cells assimilate the unknown stimuli of their phenomenology, the cortex as a composite protoplasmatic entity resembles the protoplasm of the simple amoeba which covers a body whose composite part it wants to assimilate, by turning itself into a hollow.*"[26]

Freud had referred to these ideas of Meynert's once before, albeit not linking into them positively but with a vehemently critical intent, in his study on aphasia: "*In his characteristic interpretation of anatomical relationships Meynert stated that the cortex is by virtue of its localization on the outside suitable for enveloping, taking up all sensory perceptions. He further equates it with a composite protoplasmic entity that covers a body whose composite parts it wants to assimilate by turning itself into a hollow.*" (1891*b*, p. 47f). In this way Meynert does not quite become the pioneer of a second topography, but the quote highlights the renewed virulence of anatomical concepts at the time of Freud's writing of *Beyond the Pleasure Principle* and *The Ego and the Id*, which he had studied thirty years earlier.

However, this was not the final reconciliation of psychoanalysis and neuroanatomy as is evidenced by the transformation this figure from *The Ego and the Id* has undergone in its reprinting in the *New Introductory Lectures on Psycho-Analysis* (cf. *Figure 17*). Although strikingly similar to the first version in terms of its shape, the figure nevertheless contains important changes: we are no longer dealing with a slightly distorted vesicle, but with a more abstract figure composed of several semicircular lines and open on one side. The area of the id is shown as being much smaller and the ego dominates. The open gap which indicated the area of the repressed in the first version has now been closed towards the outside and is no longer delimiting an area of the repressed against the ego. The gap itself is now the repressed, as the respective positioning of the label indicates. The 'cap of hearing'

---

[26] Theodor Meynert (1884). *Psychiatrie. Klinik der Erkrankungen des Vorderhirns begründet auf dessen Bau, Leistungen und Ernährung*. First half (a second one was never published), Vienna, p. 127. I was not able to ascertain to what extent Dorer [Dorer, M. (1932). *Historische Grundlagen der Psychoanalyse*, Leipzig] might already have discussed this reference of Freud to Meynert, since her work was not accessible to me. In his chapter about Meynert, Amacher, who cites Dorer's work, does not refer to the passage by Meynert quoted above [cf. Amacher, P. (1965). Theodor Meynert and the Anatomy of Mind. In: idem, Freud's neurological education and its influence on psychoanalytic theory, *Psychological Issues*, 4(4), Monograph 16, pp. 21–41].

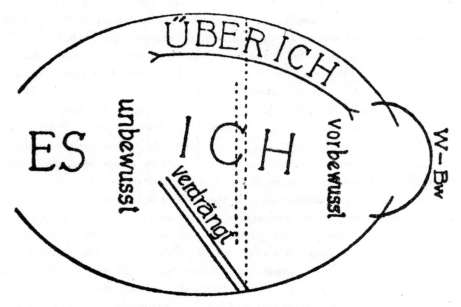

Figure 17: The psychical apparatus. In: Freud, S. (1973). Neue Folgen der Vorlesungen zur Einführung in die Psychoanalyse.
[New Introductory Lectures on Psychoanalysis]
In: Freud, S. *Gesammelte Werke*, vol. XV. Frankfurt am Main: S. Fischer Verlag, p. 85.

has been omitted and a new area, the super-ego, has been added inside the figure. One half of the labelling is tilted sideways so that there is no longer any uniform direction of looking at the figure.[27]

As a consequence of these changes all references to neuroanatomy and all similarities to a morphological representation of the brain have disappeared: the figure in the *New Introductory Lectures* purposefully provides a much more abstract schematic diagram of the psychical apparatus than did the version in *The Ego and the Id*.

---

[27] The ambivalence in the orientation of this figure has led to the fact that in the different editions of the text it has sometimes been printed upright (in the first edition and the *Studienausgabe*) and sometimes lying crosswise (in the *Gesammelte Werke*). The drafts of this figure in the manuscript of the 'New Introductory Lectures' published by Ilse Grubrich-Simitis *(Zurück zu Freuds Texten. Stumme Dokumente sprechen machen*, Frankfurt 1993, p. 201) indicate that Freud wanted to distinguish the two topographical models id/ego/super-ego and conscious/preconscious graphically by means of the different orientation of the letters. Nevertheless even the final version in the manuscript differs considerably from the printed figure. Thus e.g. a 'Vdgg' written on the side of the figure has become a 'verdrängt', in the manuscript 'ego' stretches across 'vorbewusst', between 'W-Bw' and 'vorbewusst' there are several dotted lines etc. Such differences thus also show the limitations of the present study which can only take into account the published versions of the illustrations in the *Gesammelte Werke/ StandardEdition*.

In addition, the text contains no references to a possible localization of the psychical apparatus or its parts. A comparison of these two figures of the psychical apparatus makes this later representation appear to take back all attempts at a mediation with brain anatomy that were noticeable in the earlier figure. Instead Freud generally points out the problematic nature of the visualization of the psychical apparatus when he calls this figure an "unassuming sketch"[28], remarking: "*It is certainly hard to say to-day how far the drawing is correct. In one respect it is undoubtedly not. The space occupied by the unconscious id ought to have been incomparably greater than that of the ego or the preconscious. I must ask you to correct it in our thoughts.*"[29]

The figure of the psychical apparatus in the *New Introductory Lectures* differs from that in *The Ego and the Id* not least in the fact that the id is drawn *smaller* than before. When Freud stresses in the text that this area should be imagined as greater, even "incomparably greater", the conclusion suggests itself that Freud purposefully permitted the figure to get too small in this respect, i.e. that he chose a *falsifying* graphic representation. In any case it is another illustration that includes the concept of its insufficiency, its failure as a visualization. Like the 'Anatomical schema' from the study on aphasia, this figure is an example of a rhetorical visualization that Freud uses because the psychical apparatus can only be described by language and not represented by pictures. It is at least the subsidiary goal of this representation of the psychical apparatus (in addition to illustrating the second topography) to emphasize the impossibility of visualizing the psychical with the details of this figure. It is no coincidence that Freud speaks in this context of the fact that one can do justice to "*the characteristics of the mind*" only "*by areas of colour melting into one another as they are presented by modern artists.*"[30]

This rhetorical use of illustrations is paralleled in other writings by Freud by a rhetoric of visualization in the text when he consistently describes the psychical apparatus without taking recourse to graphic representations. Thus e.g. in *A Note Upon the 'Mystic Writing-Pad'* the psychical apparatus is not described in the traditional way by a comparison with a microscope or telescope but as a writing-pad whose realistic description Freud transcends in the last step. In this way he takes the comparison from the area of the imaginable beyond it, precisely in order to do justice to the characteristic feature of the psychical.[31] Quite a similar strategy is found

---

[28] Freud, S. (1964). New Introductory Lectures. In: *Standard Edition,* vol. XXII, London: Hogarth Press, p. 78.
[29] ibid.
[30] loc.cit., p. 79
[31] For a precise reading of the "Mystic Writing-pad" see especially Jacques Derrida (1980). Freud and the Scene of Writing. In: idem, *Writing and Difference* (first published in French in 1966).

in the famous comparison with Rome in *Civilization and Its Discontent*: When developing the comparison of the psychical with a Rome, whose buildings would all have continued to exist ever since its founding days, it slowly becomes clear that the absurdity and unimaginableness of this Rome that has been preserved in all its buildings ever built is precisely the intended message. Such a Rome may be *thought of* but can no longer be *imagined*, but in order to gain clarity on this point the comparison has to be driven beyond it. The narrative quality of language allows access to the dimension of time so central to psychoanalysis for a linguistic representation. The pictorial language of the illustrations inevitably falls back behind it, but precisely in this failure it illustrates the complexity of the figures of thought. Freud's concluding remark on the comparison with Rome therefore expresses the aim of his attempts at graphic representation:

"*Our attempt seems to be an idle game. It has only one justification. It shows us how far we are from mastering the characteristics of mental life by representing them in pictorial terms.*"[32]

---

[32] Freud, S. (1961). Civilization and Its Discontent. In: *Standard Edition* vol. XXI, London: Hogarth Press, p. 71.

Erklärung der Tafel.

Fig. 1. Nervenzelle aus dem Schwanzganglion des Flusskrebses mit eingerolltem Fortsatz, welcher sich der Zellperipherie anschmiegt. Im Kern ausser den rundlichen Kernkörpern mehrere kurze, dicke Stäbchen und eine aus zwei Stücken bestehende Kernfigur. Gez. bei Hartnack 3/8. Vergrösserung der Zeichnung 360.

Fig. 2. Überlebende Nervenzelle aus einem Abdominalganglion mit kegelförmig entspringendem Fortsatz. Im Kern, welcher keine Kernmembran besitzt, vier mehrspitzige Klümpchen und ein langer, an einem Ende gebogener und gegabelter Stab. Bei k ein Kern des einhüllenden Gewebes. Dieselbe Vergrösserung.

Fig. 3. Randpartie aus dem spindelförmigen Magenganglion des Flusskrebses. Zwei unipolare Nervenzellen mit ihren Fortsätzen, deren einer eine Tförmige Theilung erfährt. Die kleinere Zelle ist bei einer Einstellung nahe der Oberfläche gezeichnet.

   s Die dicke, concentrisch geschichtete Zellscheide.
   ks Die Kerne derselben.
   hm Stark glänzende homogene Massen am Rande der Zelle, doch nach innen von der Hülle gelegen.
   f Eine von einer anderen Zelle kommende Faser.
Dieselbe Vergrösserung.

Fig. 4. Kern einer grossen Nervenzelle, welcher Bewegungserscheinungen an beiderlei Kernkörpern zeigte. b ist fünf Minuten später als a gezeichnet. Hartnack 3/X. Vergrösserung der Zeichnung 400.

Fig. 5. Stück einer Zelle mit Fortsatz wie in Fig. 1. Im Kerne eine grosse Anzahl von zierlichen gegabelten und geknickten Stäbchen. Dieselbe Vergrösserung wie in Fig. 4.

Morris N. Eagle

Freud's Legacy
Defenses, Somatic Symptoms and Neurophysiology

Introduction

I think a most fateful and fortunate step in Freud's intellectual life was his decision to abandon grand speculative neurological schemes following the *Project for a Scientific Psychology* and to limit his writings to psychological and to ontologically neutral concepts and constructs. The decision to abandon what might be called proto-neurology enabled Freud to describe and formulate psychological phenomena and principles and not especially concern himself with their neurophysiological or biological underpinnings. If the former were sufficiently valid and clear, the latter would ultimately be uncovered. This remains a useful and constructive strategy.

Freud's (1950 [1895]) *Project* is primarily of interest because of its anticipation and proto-neurological versions of his central psychological, i.e., psychoanalytical ideas. I believe a similar relationship holds between Freud's early neurological writings and his psychoanalytic writings. That is, apart from the solid contributions they may have made to such topics as aphasia, localization, and infantile cerebral paralyses, these writings are mainly of a more enduring and special interest to the extent that they suggest precursors and hints of Freud's psychoanalytic thinking. It is the latter, after all, that constitutes the main claim that Freud's ideas have on posterity.

Repressive Style

In this paper I will describe a body of research work that, I believe, constitutes an impressive instance of a heuristic and fruitful relationship between psychoanalysis and neurophysiology. Freud (1914) referred to repression as the "cornerstone" (p. 16) of psychoanalysis. It is a cornerstone both historically and conceptually. The introduction of the concept of repression marked the shift from the pre-psychoanalytic perspective of Janet, Binet, and Charcot to the birth of psychoanalysis and a psychoanalytic perspective. A core assumption made by Freud, – it is at the center of the classical psychoanalytic conception of pathology and approach to treatment, – is that both repression and lifting of repression have important consequences for psychic and somatic functioning. One hundred years later important evidence supporting this core idea has begun to accumulate. I will describe some of this evidence. The research I will discuss deals with the relationship between the use of repressive – avoidant defenses and (1) the physiological costs entailed by these defenses, including susceptibility to certain somatic illnesses, and (2) degree of cerebral laterality. I will also briefly describe work on the health benefits of writing and talking about traumatic events. These researches suggest a fruitful interaction between psychoanalysis and other related disciplines.

As we shall see, in the research I will describe, an attempt was made to understand the concept of repression and measure it in an accessible and meaningful way. Rather than attempting to demonstrate repression in the experimental laboratory, which led to fifty years of watered down and clinically meaningless analogues of repression, research on repression was revitalized with a relatively simple shift in research strategy, consisting of three components: one, treating the tendency to employ repression as a personality variable and studying individual differences in "repressive style"; two, devising means of measuring repressive style in a relatively accessible way; and three, relating individual differences in repressive style to other aspects of an individual's life which are important and central, rather than trivial. This last component bestows what one might call "ecological validity" on measures of repressive style. There are important research lessons to be learned here: one is that not every phenomenon that exists in the world can necessarily be demonstrated in an experimental laboratory. Another is that the phenomena are more important than the method and that the prime requirement of any method is that it not deform or trivialize the phenomena being investigated.

From the very introduction of the concept and throughout his writings, Freud made clear his observation and belief that repression both exacted a cost and bestowed certain benefits. Thus, early on, he writes that banishing an unwanted mental content from consciousness both frees the ego from "incompatibilities," – a benefit, – and produces hysterical conversion symptoms – a cost. Furthermore, in his later writing, Freud (1915) also refers to the work of repression, – "a persistent expenditure of force" (p. 151), – and suggests that repression entails continual psychic effort and exacts a cost on the personality. In what follows, I want to describe a body of work dealing with repressive style which strongly supports and elaborates Freud's claims regarding the costs and benefits of repression. I will also briefly describe some findings and speculations that point to some possible neurophysiological processes underlying repression.

Weinberger and his colleagues (Weinberger, Schwartz and Davidson, 1979) developed a measure of "repressive style" which has led to some fascinating results. People defined as employing a repressive coping style show the following characteristics: (1) they report low anxiety on self-report anxiety questionnaires; (2) they score high on defensiveness, as measured, for example by the Crowne-Marlowe (1964) Social Desirability Scale (Crowne and Marlowe, 1960) – that is, they deny having any unpleasant or anti-social thoughts; and (3) despite reporting low anxiety, they show a high level of physiological arousal, particularly under conditions of stress (Aspendorf and Scherer, 1983). In other words, these individuals' self-reports of a low level of anxiety are belied by their high level of physiological arousal.[1] (Weinberger, 1990). As Davidson (1983) puts it, individuals characterized by a repressive coping style "exhibit dissociations between their verbal report about how they feel and other indices, for example autonomic measures, which reflect their emotional state" (p. 351).

Benefits and Costs of Repressive Style

Subjects with a repressive defensive style tend to deny or inhibit the perception of threat and the experience of negative affect, and tend to view themselves, at least as far as conscious self-report is concerned, in a favorable light. After reviewing findings in this area, Tomarken and Davidson (1994) have characterized the repressor as showing *"a self-serving attributional style, ... a hindsight bias ..., impaired memory for negative self-relevant feedback ... and for negatively toned autobiographical events ..., attentional avoidance of threatening cues ... a relative inability to consciously perceive negative affective stimuli under specific conditions ..."* (p. 340) It may well be that those with a repressive style are the same people described by Taylor and her colleagues (e.g., Taylor, 1991) who are both somewhat self-deceptive (in a positive direction) in their self-appraisals and at a decreased risk for depression. Lane et al (1990) report that an *increased* tendency to a repressive defensive style is *inversely* related to prevalence of psychiatric disorder. In short, it appears that the use of a repressive defensive style may serve as a protective factor in regard to the experience of anxiety and depression and the psychiatric disturbances associated with anxiety and depression.

One possible explanation for the above relatively beneficial effects of repression is that it reflects the fact that the defenses of the repressive style subjects studied are intact and operate successfully in warding off anxiety, depression, and other kinds of distress associated with psychiatric disturbance. It is just what one would expect,

---

[1] Some additional possible evidence suggesting that a great deal more of an affective nature is going on internally than is reported or consciously experienced by repressors is provided in a recent study by Bell and Cook (unpublished manuscript), who report that although repressors deny having aggressive (and other 'improper') thoughts and feelings in a questionnaire, their descriptions of their recent dreams contain significantly more aggressive content than the reports of nonrepressors. Bell and Cook's interpretation of their results supports the idea that the contents repressed by repressors remain 'active' and make their presence and influence known during sleep, when repressive defenses are weakened.

based on Freudian theory of the nature and function of repression and other defenses and their relationship to anxiety. That is, when repression and other defenses operate effectively and, so to speak, do what they are supposed to do, thoughts and affects that would produce dysphoric affects and that would be disturbing to one's self esteem would tend to be kept from conscious awareness. And, as one can see from Tomarken and Davidson's (1994) description, this is just the way repressors function.

I am reminded of a finding in another area, – attachment research, – which seems quite related to work on repressive style. Adams, Keller & West (1995) reported that subjects with an *avoidant* attachment style – a personality characteristic that would appear to bear important similarities to repressive style – are at a *decreased* risk (compared to those with an enmeshed/ preoccupied attachment style) for suicide and suicide attempts. In other words, being avoidant serves as a protective factor against suicide attempts, – and presumably, against the experience of the intense distress associated with suicide attempts.

The story now gets more complicated. Yes, a repressive style appears to protect one against the experience of anxiety, depression, lowered self-esteem, and other dysphoric affects. However, recall the finding that repressive style subjects show a heightened level of autonomic arousal. There is evidence suggesting that they also show a decreased immune response under stress (Schwartz, 1990), heightened systolic blood pressure (King et al, 1990), increased salivary cortisol levels (Brown et al, 1993), and a heightened susceptibility to a variety of somatic conditions including ulcers, allergies, hypertension, impotence and vaginal herpes (Schwartz, 1990).

Jensen (1987) found that repressor cancer patients reported significantly less negative affect and fewer negative side effects of treatment (such as nausea, pain, and sleep disturbance [Ward, Leventhal, and Love, 1988, as reported in Jensen, 1987]). However, at a two year follow-up of breast cancer patients, repressors "had significantly shorter periods of remission and were significantly more likely at follow-up to show metastasis, medical deterioration and death" (Jensen, 1987, p. 338). This pattern of short-term benefit and long-term harm associated with the repressive style was also found for coronary patients. Schwartz (1990) reported that coronary patients scoring high on the Levine Denial of Illness Scale (which, one assumes, would be highly correlated with repressive style) showed less cardiac dysfunction during the acute period of their illness but fared poorly after leaving the hospital (see also Levine et al, 1987). Repressors also show poorer control of the latent Epstein-Barr Virus as evidenced by high titers for the viral capsid antigen (Esterling et al, 1990). In other studies, repressors show lower monocyte count, greater serum glucose levels, and report significantly greater allergic reactions to opiate medication (Jamner, Schwartz, & Leigh, 1988).

What conclusions can one draw from the above (plus other related) work? It appears that the use of repressive defenses and a repressive coping style serves to protect one against the conscious experience of negative affect, negatively toned memories, and in general, information that is threatening to one's self-esteem. That is, so to speak, the good news. The bad news is that a somatic cost is exacted in the form of heightened autonomic activity and the physical conditions (e.g., hypertension, compromised immune response) that are associated with such heightened activity. This pattern is consistent with the finding that inhibition of negative affect is associated with heightened autonomic activity (Gross and Levenson, 1993; Koriat et al, 1972).

Thus, the research on "repressive style" provides impressive support for Freud's basic claim that although repression bestows certain advantages, the work of repression exacts a cost. Freud (1937) writes, "*It sometimes turns out that the ego has paid too high a price for the services they [i.e., defense mechanisms] render it. The dynamic expenditure necessary for maintaining them, and the restrictions of the ego which they almost invariably entail, prove a heavy burden on the psychical economy*" (p. 237). The recent work on repression that I have briefly summarized suggests that repression may entail a heavy burden, not only on the psychical economy, but also on the somatic economy.

One may question whether the repression of "repressive style" really captures what Freud meant by repression. After all, in classical theory, repression refers to the banishment of drive derivatives from consciousness. The fact is, however, that both prior to the development of his drive theory and scattered throughout his writings, Freud did not limit the application of the concept of repression to instinctual wishes, but employed it in a much broader context, to refer to any mental contents inimical to or sharply incompatible with the ego and one's self-image. However, quite apart from the precise meaning that Freud intended, if one understands the essence of repression to be the exclusion from consciousness of emotionally distressing mental contents (i.e., thoughts, desires, wishes, fantasies, etc.), then the current work on repressive style seems to represent an important contribution to a fuller understanding of the concept of repression, including such issues as the benefits it bestows, the cost it exacts, and the contexts and circumstances that influence its relative benefits and costs.

My view is that as one increasingly attempts a systematic empirical, but ecologically valid, investigation of psychoanalytic ideas, propositions and concepts, they will not necessarily survive in their original "pure" and unaltered form. The very attempt to reliably measure psychoanalytic concepts will demystify them and we will be left with 'something like' the original concepts. In addition, new concepts and formulations that are linked to general cognitive and affective processes will be generated. These are not developments to be lamented as compromising the purity of psychoanalysis, but rather important and legitimate attempts at demystification and integration with other fields to be encouraged.

Repressive Style and Cerebral Laterality

Up to this point, I have been discussing the physiological and psychological costs and benefits of repression. Let me turn now to recent theoretical speculations and findings concerning the neurophysiological processes that may underlie repression and other defenses. There are two main components to work in this area that I will briefly review. (1) There is increasing evidence that contrary to an early hypothesis, the right hemisphere is not the "seat" of all emotion, but rather is particularly specialized for at least certain negative emotions, while the left hemisphere is specialized for the processing of at least certain positive emotions as well as verbal report and self relevant feedback; (e.g., Davidson,, 1983; 1984; Dimond, Farrington, & Johnson, 1976. (2) A number of investigators have speculated and have presented evidence suggesting that repression and repressive coping style may be associated with a functional deconnection between left hemispheric processing involved in verbal self-concepts and report of feeling states on the one hand, and right hemisphere processes which predominate in autonomic and neuroendocrine responses to negative affective stimuli, on the other. (See Galin [1974], who, early on, speculated that repression and other ego defenses may entail a degree of dissociation or functional deconnection between the cerebral hemispheres.)

There is accumulating evidence supporting these speculations. Let me briefly note some of these findings. Some of the observations and findings that have led to the first hypothesis, namely, differential hemisphere specialization of negative and positive affects include the following: Many years ago, Kurt Goldstein (1939), – one of my teachers, – was one of the first to report a high incidence of negative affect and "catastrophic" reactions among patients with unilateral left-hemisphere damage. In contrast, indifferent or euphoric reactions, the latter characterized by inappropriate laughter and joking, is frequently reported to follow unilateral right-hemisphere damage. Left hemisphere frontal damage is associated with less spontaneous smiling and greater depression. Depressed subjects show greater right-frontal EEG alpha activation than non-depressed subjects. Right-sided electro-convulsive therapy (ECT) has been reported to be more effective for depression than left-sided unilateral ECT, possibly suggesting that the 'seat' of depression is more likely to be the right hemisphere. (See Davidson, 1983; 1984 for a review of this literature).

Questions that are likely to produce negative affects elicit a greater number of left and a lesser number of right eye movements. Subjects reported experiencing more happiness in response to faces presented to the right visual field and, therefore, to the left hemisphere. EEG studies revealed greater right hemisphere activation when sad memories were hypnotically revivified compared to happy memories. The same pattern was found with hypnotically induced moods (see Davidson, 1983; 1984). In infants as young as 10 months of age, observing happy epochs in which a person's neutral face changes into a smiling and laughing face, is associated with greater left frontal activation than observing sad epochs, characterized by the neutral face changing into a frowning and crying face (Davidson, & Fox, 1982). When films were shown to adult subjects' right or left visual field, they rated them as more unpleasant and horrific when shown to the right hemisphere (Dimond, Farrington, & Johnson, 1976). And, happy faces are associated with faster reaction time when presented to the left hemisphere. Sad faces produce the opposite results. As a final example of the evidence supporting the hypothesis that the right and left hemispheres are specialized, respectively, for certain negative and positive affects although one finds, on average, a faster and more accurate recognition of all emotional faces when they are presented to the right as compared with the left hemisphere, sad faces produce the greatest right hemisphere advantage and happy faces produce the greatest number of errors when presented to the right hemisphere. (See Davidson, 1983).

There is evidence supporting the second hypothesis, that repressors show a functional deconnection between the two hemispheres characterized by deficits in the transfer of negative affective information processed by the right hemisphere to the left hemisphere. For example, subjects were presented with affective faces exposed tachistoscopically either to the left or right visual field and were asked to name the emotion depicted in the face. When the faces were presented to the left hemisphere, there were no differences in accuracy between repressors and non-repressors. However, when the faces were presented to the right hemisphere, the repressors were significantly less accurate than the non-repressors, suggesting a relative deficit in transfer of affective information from the right to the left hemisphere (see Davidson, 1984). As another example, Gur and Gur (1975) found that subjects with right hemisphericity, as inferred from the dominant direction of conjugate lateral eye movements, employed a repressive style, as determined from scores on the Defense Mechanism Inventory. These findings support the general idea that the hemisphericity and type of defense one employs are related to each other. As Davidson, (1983, 1984) notes, the overall results are compatible with the hypothesis that those employing a repressive style show relative deficits in interhemispheric transfer of negative affective information. Such a deficit may underlie the frequently reported dissociation between verbal report (e.g., of anxiety) and non-verbal indices of affective responding, for example, autonomic measures characterizing repressors.

Why Does Repressive Style Lead to Somatic Symptoms?

If one puts together the various findings I have described, the following picture of repressors emerges: It appears that repressors do process and respond to affective information but do not become consciously aware of the feeling states and self implications normally associated with this negative affective information. Instead, they respond on the autonomic and neuroendocrine level. In remarkable congruence with Freud's early formulations, one finds that repressors have spared themselves the experience of, to quote Freud, "an incompatibility ... between the ego and some idea presented to it," – as reflected in their report of low anxiety, denial of antisocial thoughts, and higher self-esteem, – but pay the price in somatic symptoms and illnesses.

Freud accounted for the relation between repression and somatic symptoms in hysteria by speculating that the distressing affect that was not consciously experienced, – the so-called "strangulated affect" (Freud, 1893, p. 39) – was directly *converted* into somatic symptoms (Freud, 1894, p. 49)- hence, the term "conversion symptoms." Current work on repressive style suggests that repression is associated with greater susceptibility to a wide range of somatic symptoms and organic illnesses, not simply hysterical conversion symptoms. How does one account for this relation?

Some current responses to this question are compatible with and constitute elaborations of a version of Freud's early concept of strangulated affect and the idea that repression entails work and a "persistent expenditure of force" (Freud, 1915, p. 151). Thus, Esterling et al (1990) hypothesize "that to inhibit ongoing behavior, thoughts, and feelings, physiological work must be performed." They go on to note that "immediately following a stressful event, inhibition is accompanied by increases in autonomic nervous activity and by increased sympathetic firing in the septal and hippocampal regions of the brain" and conclude that "it is plausible to propose that if this behavioral pattern continues, the work of inhibition may serve as a low level chronic stressor which has the potential of long-term cumulative damage" (p. 407). Weinberger (1990) describes this aspect of repressive style as "analogous to a car that maintains a low speed because the driver is leaning on both the breaks and the accelerator" (p. 372). In accordance with this view, Schore (1994) speculates that a "chronic stressful state of [both] heightened excitation and heightened inhibition ... may characterize the psychobiology of repression" (p. 325).

Another reason that the use of repression may be related to somatic symptoms is that because repressors do not become aware of negative affective information, they fail to reflect on this information, on the implications such information might have for their life, and therefore do not do anything or attempt to work through the issues that are raised by negative affective information and states. Instead, these continued negative affective states result in chronic autonomic activation and other physiological responses, including heightened release of stress-related hormones such as cortisol and epinephrine, heightened systolic blood pressure, and suppression of the immune system – and eventually, increased susceptibility to certain somatic diseases.

During a period in psychoanalytic history in which an emphasis on corrective emotional experiences has increasingly overshadowed the value of insight and self-awareness, it is important to be reminded of the rather commonsensical point that becoming aware of one's patterns, attitudes, feelings, thoughts, scripts, and so on is frequently a necessary (even if not always sufficient) step in altering those patterns, attitudes, feelings, etc. As Weinberger (1990) notes, one of the ways we have of coping with a stressful situation is to try to modulate our emotions by altering our initial affective appraisal – e.g., reasoning why the situation should not be that upsetting. However, if "the repressive mechanism entails ignoring the emotion system rather than informing it" (p. 366) and if "the affective appraisal becomes as much a threat as the external situation," then such coping strategies are not available. To quote Weinberger (1990) again, "Whistling in the dark or various cognitive equivalents are not very effective in convincing the emotion system that its reactions are unwarranted." The relative absence of affective appraisal can, as Schwartz (1983) notes, lead to disregulated subsystems functioning without the stabilizing effects of appropriate feedback.

Another mechanism that may account for the deleterious consequences of repression, particularly in regard to physical illness, is that it may interfere with appropriate health behavior. As noted earlier, although deniers[2] do well during the acute phase following a heart attack, they do more poorly during the six month follow up period (Levine et al, 1987; Schwartz, 1990). A plausible explanation for the latter is that because of their denial they do not implement necessary life style changes. In support of this explanation, Shaw et al (1985) reported that as a group repressors who had suffered a heart attack retained less information about lifestyle risk factors than non-repressors during a rehabilitation program. Insofar as they tend to shut out or minimize internal cues, repressors may also be less aware of and less attentive to symptoms that would warrant medical consultation and treatment.

Still another factor in the deleterious consequences of repressive defenses has to do with what in psychoanalytic theory one would call the harshness of the superego. As Weinberger (1990) makes clear, repressors tend to avoid acknowledging and experiencing their "egoistic" affects and impulses that are incompatible with their over-idealized self-image. In this regard, they make demands on themselves that are not in accord with the feelings, impulses and desires of ordinary mortals. It is inter-

---

[2] I am assuming that deniers are likely to employ a repressive style.

esting to note that in his later writings, Freud (1926 [1925]) observes that repression is instigated at the "behest of the superego" (p. 91) and an 'ego-ideal' set of standards that are at odds with one's thoughts, feelings, etc. With regard to the latter, he writes: "...the formation of an ideal heightens the demands of the ego and is the most powerful factor favoring repression" (Freud, 1914, p. 95). In other words, repression is triggered by a harsh superego that will not accept feelings and thoughts that do not live up to one's ego-ideal, that is, that do not live up to an idealized self-image. And, indeed, Weinberger (1990) summarizes a set of evidence that is entirely in accord with this formulation. For example, sixth-grade repressors tend to be described by their classmates as showing high restraint, less prone to misconduct, and high in self-control (p. 370). In different samples of college students, repressors were low in both alcohol and illegal drug use (p. 371). As compared with non-repressors, repressors are also significantly more tolerant of pain.

If one were to try to put together the work on repression, and repressive coping style, – and I have presented only a modest sample of this work in this chapter, – one would, I believe, need to set up a master balance sheet that would include, not only a summary of the general costs and benefits of repressive defenses and coping style, *but would also try to relate these costs and benefits to the specific contexts*[3] of the individual. One would need to do the same regarding the costs and benefits of lifting repressive defenses. This is what one informally and intuitively tries to do in good clinical work.

The Beneficial Effects of Moderating Repression

If a repressive style is associated with risk for certain somatic illnesses, one would expect that at least under certain circumstances, something akin to lifting repression might have a positive impact on physical health. And indeed, there is a body of recent work indicating that for a large group of people, thinking, writing, and talking about stressful, disturbing and traumatic events seems to have a significant positive impact on their general somatic health (e.g., Pennebaker, 1993; 1995; 1997), on the pattern of specific illnesses and on specific physiological functions. For example, Pennebaker and his colleagues have consistently found that among college students writing about a traumatic event is associated with fewer visits to the student health service (e.g., Pennebaker, & Beall, 1986; Pennebaker, Colder & Sharp, 1990). As another example, Esterling and his colleagues (1990; 1994) have reported that high emotional disclo-

---

[3] One would, I believe, need to do the same with the costs and benefits of other defenses and coping styles, including the so-called sensitizing coping style, which, in important respects, is presumably the opposite of the repressive coping style. There is little doubt that the perceptual vigilance and other features of the sensitizing style also both bestow certain benefits and exact certain costs, and that the costs and benefits of this coping style are also context dependent.

sure of traumatic events through writing or speaking modulates the latent Epstein-Barr Virus and is associated with lower levels of immunoglobulin anti-body. As still another example, Pennebaker, Kiecolt-Glaser & Glaser (1988) found that disclosure of trauma was associated with improved immune function. And as a final example, high disclosures show greater decline in systolic and diasystolic blood pressure than low disclosures. (Pennebaker, Hughes, O'Heeron, 1987)

The Continuing Vitality of the Concept of Repression

There is a certain irony about the work in this area. During a period in psychoanalytic history in which we are repeatedly told by some critics that virtually all Freudian concepts and formulations are eminently discardable, researchers on repressive style are quite clear regarding not only their debt to Freudian theory, but also their belief in the continuing validity of many aspects of these concepts and formulations. Thus, at the very same time that the concept of repression – which Freud (1914) viewed as the "cornerstone" (p. 16) of psychoanalysis – is virtually absent from contemporary psychoanalytic theories – it is alive and well in contemporary research inspired by psychoanalytic theory. Also, the broad way of understanding repressive style reflected in this research – as the use of avoidant psychological strategies to keep from conscious awareness certain mental contents that are distressing because they are sharply discrepant with one's self-image – is virtually identical to Freud's conceptualization when he first introduced the concept and stressed "an incompatibility ... between the ego and some idea presented to it" (Breuer & Freud, 1893–1895, p. 122) as the essential stimulus for repression. During this early period, before Freud developed his drive theory, there was no talk about the special susceptibility of infantile instinctual impulses to repression. The focus was on any significant incompatibility between the ego and a mental content presented to it. It seems to me that the current work on repressive style suggests that Freud's earliest formulation of repression remains the most useful and most valid one.

Personal Meanings and Physiological Mechanisms

The introduction of the concept of repression marked the emergence of a distinctive psychoanalytic voice and perspective. According to that perspective, the failure to integrate and admit certain mental contents to consciousness is not attributable to a constitutional "psychical insufficiency" (Breuer & Freud, 1893–1895, p. 104) in one's capacity for psychical synthesis, or as a "form of degeneracy" (Freud, 1894, p. 51, fn. 3) as Janet maintained, but is the result of a purposive and motivated act. Once repression was introduced, personal meanings and motives became the predominant way of understanding psychopathology and, indeed, much of human behavior. This shift from constitutional weakness to repression is but one expression of a historical and chronic antagonism between a perspective in which human behavior is accounted for in terms of personal meanings and motives and a perspective in which human behavior is viewed as the automatic causal consequence of, so to speak, 'meaningless' processes (see Gill, 1976; also see Eagle, 1980; 1987). Although Freud opted for the perspective of personal meanings and motives, throughout his metapsychological writings he introduced impersonal concepts, forces, entities, and trends such as mental apparatus, libido, psychic energy, pleasure principle, and so on. Most important, he was always aware of the need to integrate the two perspectives of what Holt (1972) refers to as the humanistic and scientific images of man.

It seems to me that the work on repressive style represents a partial implementation of, to borrow a term from Lakatos (1970), Freud's 'research program'. One can understand repressive style in terms of personal meanings and motives – for example, the need to maintain a certain self-image or the need to ward off negative affect. However, a deep and comprehensive explanatory account does not end there. One wants to investigate the neurophysiological underpinnings and mechanisms that underlie these personal meanings and motives. For example, precisely how do the goals of avoiding the conscious experience of negative affect and of thoughts that would disturb one's self-image get implemented? Also, how does the implementation of these goals lead to somatic symptoms and illnesses? What are the processes involved?

The work on repressive style that I have described begins to look at and attempts to answer these questions. Most important from a methodological and philosophical point of view, this work suggests that one need not dichotomize between personal meanings and motives on the one hand and biological processes and mechanisms on the other. As Rubenstein (1997) noted, not only are we both persons and organisms, but the pursuit of personal meanings and motives is: (1) ultimately generated, at least in part, by biological processes; (2) is always implemented by biological processes; and (3) always has biological consequences. As Freud (1950 [1895]) was aware since the *Project*, it is not useful to reduce one to the other. Rather, one should study each at its own level of integrity as well as the complex relations between the proximal psychological level and the distal biological one. In no other psychology or psychoanalytic theory is this insight as deeply embedded as in Freudian theory.

REFERENCES

ADAMS, K. S., KELLER, A. E. S., & WEST, M. (1995). Attachment organization and vulnerability to loss, separation, and abuse in disturbed adolescents. In: S. Goldberg, R. Muir, & J. Kerr (Eds.) *Attachment theory: Social, developmental, & clinical perspectives.* Hillsdale, New Jersey: Analytic Press, pp. 309–341.

ASPENDORF, J. & SCHERER, K. (1983). The discrepant repressor: Differentiation between low anxiety, high anxiety, and repression of anxiety by autonomic-facial-verbal patterns of behavior. *Journal of Personality and Social Psychology*, 44, 133–147.

BELL, A. & COOK, H. (Unpublished manuscript). Empirical evidence for a compensatory relationship between dream content and repression.

BREUER, J. & FREUD, S. (1893–1895). *Studies on hysteria. Standard Edition*, vol. II., London: Hogarth Press, 1955, pp. 3–307.

BROWN, W. A., SIROTA, A. D., NIAURA, R., ENGEBRETSON, T. O. (1993). Endocrine correlates of sadness and elation. *Psychosomatic Medicine*, 55 (5), 458–467.

CROWNE, D. P. & MARLOWE, D. (1964). *The approval motive: Studies in evaluative dependence*, New York: Wiley.

DAVIDSON, R. J. (1983). Affect, cognition, and hemispheric specialization. In: C. E. Izard, J. Kagan, R. E. Zajonc (Eds.) *Emotions, cognitions and behavior*, Chicago: University of Chicago Press, pp. 387–403.

DAVIDSON, R. J. (1984). Hemispheric asymmetry and emotion. In: K. Scherer & P. Ekman (Eds.) *Approaches to emotion*, Hillsdale, NJ: Lawrence Erlbaum Associates, pp. 39–57.

DAVIDSON, R. J. & FOX, N. A. (1982). Asymmetrical brain activity discriminates between positive versus negative affective stimuli in ten month old infants. *Science*, 218, 1235–1237.

DIMOND, S. J. & FARRINGTON, L., & JOHNSON, P. (1976). Differing emotional response from right and left hemispheres. *Nature*, 261, 690–692.

EAGLE, M. N. (1980). A critical examination of motivational explanation in psychoanalysis. *Psychoanalysis and Contemporary Thought*, 3, 329–380. Also in: L. Laudan (Ed.) *Mind and Medicine: Problems of explanation and evaluation in psychiatry and the biomedical sciences*. University of Pittsburg series in philosophy and history of science, Los Angeles & Berkeley, California: University of California Press, pp. 311–353.

EAGLE, M. N. (1987). *Recent developments in psychoanalysis: A critical evaluation*. Cambridge, MA: Harvard University Press.

ESTERLING, B. A., ANTONI, M. H., KUMAR, M., & SCHNEIDERMAN, N. (1990). Emotional repression, stress disclosure responses, and Epstein-Barr viral capsid antigen titers. *Psychosomatic Medicine*, 52, 397–410.

FREUD, S. (1893). On the psychical mechanism of hysterical phenomena: A lecture. In: *Standard Edition*, vol. III, London: Hogarth Press, 1962, pp. 25–39.

FREUD, S. (1894). The neuro-psychoses of defense. In: *Standard Edition*, vol III, London: Hogarth Press, 1962, pp. 41–61.

FREUD, S. (1914). On narcissism: An introduction. In: *Standard Edition*, vol. XIV, London: Hogarth Press, 1957, pp. 68–102.

FREUD, S. (1914). On the history of the psychoanalytic movement. In: *Standard Edition*, vol. XIV, London: Hogarth Press, 1957, pp. 2–66.

FREUD, S. (1915). Repression. In: *Standard Edition*, vol. XIV, London: Hogarth Press, 1957, pp. 141–158.

FREUD, S. (1926 [1925]). Inhibition, symptoms & anxiety. In: *Standard Edition*, vol. XX, London: Hogarth Press, 1959, pp. 75–174.

FREUD, S. (1937). Analysis terminable and interminable. In: *Standard Edition*, vol. XIV, London: Hogarth Press, 1957, pp. 209–259.

FREUD, S. (1950 [1895]). Project for a scientific psychology. In: *Standard Edition*, vol. I, London: Hogarth Press, 1966, pp. 283–397.

GALIN, D. (1974). Implications for psychiatry of left and right cerebral specialization. *Archives of General Psychiatry*, 26, 15–23.

GILL, M. M. Metapsychology is not psychology. (1976). In M. M. Gill & P. S. Holzman (Eds.) Psychology versus metapsychology: Psychoanalytic essays in memory of George S. Klein. *Psychological Issues*, 9 (Monograph No. 36), 71–105.

GOLDSTEIN, K. (1939). *The Organism: A Holistic approach to biology derived from pathological data in man*, New York: American Books.

GROSS, J. J. & LEVENSON, R. W. (1993). Emotional suppression: physiology, self-report, and expressive behavior. *Journal of Personality and Social Psychology*, 64, 970–986.

GUR, R. E. & GUR, E. C. (1975). Defense mechanisms, psychosomatic symptomatology, and conjugate lateral eye movements. *Journal of Consulting and Clinical Psychology*, 43, 416–420.

HOLT, R. R. (1972). Freud's mechanistic and humanistic images of man. *Psychoanalysis and Contemporary Science*, 1, 3–24.

JAMNER, L. D., SCHWARTZ, G. E., & LEIGH, H. (1988) The relationship between repressive and defensive coping styles and monocyte, eosinophile, and serum glucose levels: Support for opioid-peptide hypothesis of repression. *Psychosomatic Medicine*, 50, 567–575.

JENSEN, M. R. (1987). Psychobiological factors predicting the course of breast cancer. *Journal of Personality*, 55, 317–342.

KING, L. A. & EMMONS, R. A. (1990). Conflict over emotional expression: psychological and physical correlates. *Journal of Personality and Social Psychology*, 58 (5), 864–877.

KORIAT, A., MELMAN, R., AVERILL, J. R., & LAZARUS, R. S. (1972). The self-control of emotional reactions to a stressful film. *Journal of Personality*, 40, 601–619.

LAKATOS, I. (1970). Falsification and the methodology of scientific research programs. In: I. Lakatos & A. Musgrave (Eds). *Criticism and the growth of knowledge*, New York: Cambridge University Press, pp. 91–195.

LANE, R. D., MERIKANGAS, K. R., SCHWARTZ, G. E., HUANG, S. S. (1990). Inverse relationship between defensiveness and lifetime prevalence of psychiatric disorder. *American Journal of Psychiatry*, 147 (5), 573–578.

LEVINE, J., WARRENBURG, S., KERNS, R., SCHWARTZ, G. E., DELANEY, R., SONTANA, A., GRADMAN, A., SMITH, S., ALLEN, S., CASCIONE, R. (1987). The role of denial in recovery from coronary heart disease. *Psychosomatic Medicine*, 49 (2) 109–117.

PENNEBAKER, J. W. (1993). Putting stress into words: Health, linguistic, and therapeutic implications. *Behavior Research and Therapy*, 31, 539–548.

PENNEBAKER, J. W. (1995). *Emotion, disclosure and health* Washington, DC: American Psychological Association Books.

PENNEBAKER, J. W. (1997). *Opening up: The healing power of expressing emotion*, New York: Guilford Press.

PENNEBAKER, J. W., & BEALL, S. K. (1986). Confronting a traumatic event: Toward an understanding of inhibition and disease. *Journal of Abnormal Psychology*, 95, 274–281.

PENNEBAKER, J. W., COLDER, M., & SHARP, L. K. (1990). Accelerating the coping process. *Journal of Personality and Social Psychology*, 58 (3), 528–537.

PENNEBAKER, J. W., HUGHES, C. F., & O'HEERON, R. C. (1987). The psychophysiology of confession: Linking inhibitory and psychosomatic processes. *Journal of Personality and Social Psychology*, 52 (4), 781–793.

PENNEBAKER, J. W., KIECOLT-GLASER, J. K., & GLASER, R. (1988). Disclosure of traumas and immune function: Health implications for psychotherapy. *Journal of Consulting and Clinical Psychology,* 56 (No. 2), 239–245.

RUBINSTEIN, B. (1997). Person, organism, and self: Their worlds and their psychoanalytically relevant relationships. In: R. R. Holt (Ed.) *Psychoanalysis and the philosphy of science. Collected papers of Benjamin B. Rubinstein., M. D.* Psychological Issues Monograph 62/63, Madison, CT: International Universities Press, pp. 415–465.

SCHORE, A. N. (1994). *Affect regulation and the origin of the self: The neurobiology of emotional development,* Hillsdale, NJ: Lawrence Erlbaum Associates.

SCHWARTZ, G. E. (1983). Disregulation theory and disease: Applications to the repression/ cerebral disconnection/ cardiovascular disorder hypothesis. *International Review of Applied Psychology,* 32 (2), 95–118.

SCHWARTZ, G. E. (1990). The psychobiology of repression and health. In: J. Singer (Ed.) *Repression and dissociation,* Chicago: University of Chicago Press.

SHAW, R. E., COHEN, F., DOYLE, B., PALESKY, J. (1985). The impact of denial and repressive style on information gain and rehabilitation outcomes in myocardial infarction patients. *Psychosomatic Medicine,* 47 (3), 262–273.

TAYLOR, S. E., HELGESON, V. S., REED, G. M., SKOKAN, L. A. (1991). Self-generated feelings of control and adjustment to physical illness. *Journal of Social Issues,* 47 (4), 91–109.

TOMARKEN, A. J. & DAVIDSON, R. J. (1994). Frontal brain activation in repressors and nonrepressors. *Journal of Abnormal Psychology,* 103 (2), 339–349.

WARD, S. E., LEVENTHAL, H., LOVE, R. (1988) Repression revisited: Tactics in coping with severe health threat. *Personality and Social Psychology Bulletin,* 14 (4), 735–746.

WEINBERGER, D. R. (1990). The construct validity of the repressive coping style. In: J. Singer (Ed.) *Repression and dissociation,* Chicago: University of Chicago Press.

WEINBERGER, D. R., BERMAN, K. F., ZEC, R. F. (1986). Physiologic dysfunction of dorsolateral prefrontal cortex in schizophrenia: I. Regional blood flow evidence. *Archives of General Psychiatry,* 43 (2), 114–124.

WEINBERGER, D. R., SCHWARTZ, G. & DAVIDSON, R.. (1979). Low-anxious, high-anxious, and repressive coping styles: Psychometric patterns and behavioral and physiological responses to stress. *Journal of Abnormal Psychology,* 88, 369–380.

Erklärung der Abbildungen.

Fig. 1. Die Hälfte eines Querschnittes des Rückenmarks von *Ammocoetes*, aus Müller'scher Flüssigkeit. Ein Stück der vorderen, äusseren Ecke fehlt.

c. Centralkanal,
h. Hinterzelle,
hzf. Hinterzellenfortsatz,
M. f. Müller'sche Faser,
v. Vorderhorn.

Fig. 2. Ein Querschnitt durch den ganzen *Ammocoetes*, Chromsäurepräparat. Die den Rückenmarkskanal umgebenden Gewebe sind nur theilweise gezeichnet.

Ch. Chorda dorsalis.
Chs. Die drei Schichten der inneren Chordascheide.
d. Dura mater.
p. Pia mater.
ar. Zellen und elastische Fasern im Arachnoidealraum.
m. Muskelsegmente.
n. l. Querschnitt des *nervus lateralis*.
M. f. Müller'sche Faser.
c. Centralkanal.
h. Hinterzelle.
h. f. Hinterzellenfaser.

Daneben andere Wurzelfasern,

f. die man nicht zu Hinterzellen verfolgen kann.
h. w. hintere Wurzel.
s. G. umgebendes fetthaltiges Gewebe, in dem bei *Petromyzon* das knorplige Skelet liegt.

Detlef B. Linke

Discharge, Reflex, Free Energy and Encoding

There is no discharge

Energetic considerations play a decisive role in the underlying concepts of Freud's psychological work. In his *Entwurf einer Psychologie [Project for a Scientific Psychology]* the importance of models of thermodynamics is quite clear. Generally Freud regards the lowest possible energy level as beneficial to the functioning of the nervous system. Discharge is seen partly in analogy to the reflex activity as a means to reduce the energy level. The idea that the energy of the psyche and the nervous system, respectively, can be discharged towards the outside has been very significant in the popular reception of psychoanalysis. It must be pointed out, however, that with regard to the reflexes the discharge metaphor lacks any physiological underpinnings. In reflexes no energy is transferred from the reflex center to the periphery. In the discharge of nerve cells energy is used up right then and there. When an action potential occurs an electromagnetic field is generated which is able to trigger potential shifts in adjacent membrane zones. These shifts in potential brought on by the movement of ions across the cell membrane in turn constitute an energy-consuming process so that the transmission of impulses in the nerve fiber consists of the activation of a concatenation of local energy-consuming processes. There is a flow of axoplasm from the perikaryon, i.e. the cell body surrounding the nucleus, to the peripheral end of the axon. This is, however, a long-lasting transport of energy that does not have any effect on short-term impulse transmission.

The axonal transfer of energy can, in many respects, be viewed as similar to what happens in the system of blood circulation: energy is made available on a certain level. In the case of energy being supplied by the circulation of blood (above all in the form of glucosis) one cannot assume any pressure to discharge. The energy demand depends on the constructive metabolism and on the actions of the organism. With sufficient nourishment and oxygen supply the influence of the available energy level upon the activities of the organism is rather negligible in terms of its general psychological balance. The significant coordination takes place in the opposite direction. The blood supply of the various regions of the brain is controlled by demand. Whenever cognitive or emotional processes in a specific region of the brain result in a certain activity of the neurons in question, vasodilation occurs in this area to provide an increased supply of blood and energy. Thus energy is normally available for cognition, emotions and "information processing" and is supplied regulatively.

Muscular activity likewise does not represent a discharge of energy by the central nervous system, but is made possible by energies made available by the blood circulation in a regulatory way according to need. The triggering of muscular activity by an arbitrary impulse in the motor centers of the brain does not constitute a discharge of energy from the center to the periphery. With regard to motor responses the discharge metaphor thus lacks any physiological basis. This does not mean, however, that the energetic view of the processes occurring in the nervous system is irrelevant to psychology, and it also does not exclude the possibility that certain psychological processes can be understood better in analogy to thermodynamic processes on the basis of the physiology of the central nervous system. It likewise does not imply that any talk of discharge has lost all meaning. However, the justification of the metaphor must be reexamined and its possible misleading elements analyzed. Retaining for the moment the idea that too much energy in the nervous system can trigger pathological states, it would then not be the discharge of this energy to the outside that could play a therapeutic role but rather its transformation according to the laws of recent neurophysiological insights. This means first of all that a new understanding of "discharge" need not imply a movement from within to without and that "discharge" need not be limited to motor actions.

Free Energy

Information and energy are linked at least to the extent that information requires an energetic signal carrier. The amount of energy used may vary. Information can be transmitted from one mountain top to another by means of the low energetic impulse of a flashlight or by an enormous fireworks that is supposed to be of significance. In the nervous system the range of the linkage between energy and information is not random. First of all, the energetic resources available in the nervous system are limited. Furthermore, if many times the usual amount of energy is expended for a specific information this may influence the processing of information on the energetic level in other areas that operate on a lower energetic level.

In terms of energy, information is a measure of the unlikeliness of a certain energy state occurring. With regard to the second law of thermodynamics, information indicates that energy-rich states are not just unlikely but that there are also possibilities of agreement that permit a differentiated coupling of energy and information.

In recent years energetic approaches have found little theoretical consideration in the study of emotions and cognitions in the neurosciences (see e.g. Freeman).

The specificities of the "logic" of the nervous system, however, seem to be related to the fact that the connections not only take into account aspects of information theory but also consider the energetic stabilization of a state. From this point of view the non-transmission of an impulse by a neuron can also be seen as an attempt by the organism to reduce its energy expenditure and to concentrate only on the most important matters. Thus the interruption of an impulse does not need to mean that it has come to a standstill in a negation or switching but it can also mean that as part of a certain "economizing" of information processing importance is being attributed to the signals in terms of significance rather than meaning.

To a certain extent information processing, the analysis of the probability of impulses occurring in the nervous system, is also responsible for providing energy levels. It would therefore be a mistake to try to understand the information configured by the signal transmission only from the point of view of information theory. On the other hand, it would not be appropriate to describe the activity of the nervous system merely in terms of signal processing. If this were the case, there would be neither the selection mechanisms nor the aspects of importance that can be characterized merely behaviorally in connecting e.g. to the motor areas, nor would there be the dimension of information that is registered by the nervous system in the sense that unlikely states lead to different reactions from likely states.

Without taking the energetic dimension into account, however, the search for the principle of information processing and for the code of the nervous system does not lead very far. The nervous system is not equipped with a clock as the ordinary computer is, but has very differently defined temporal windows for the various levels of information processing. The temporal parameters for conveying action potential in the axon, for transmission between synapses, for the break-down of transmitters, fluctuations in potential, etc. are all very different so that this fact alone makes it difficult to work with defined codes. In information processing one usually speaks of an analog pulse repetition frequency code with a certain bandwidth in the axon. Fluctuations in membrane potential are likewise partly analog, whereas the action potential is subject to a threshold logic.

Even if it were possible to define different encodings for the various parts of the nervous system, considerable difficulties are presented by the fact that the different temporal parameters of items of information that arrive simultaneously e.g. through different media systems, bear their own temporal characteristics and cannot easily lead to a balance of information processing in the temporal dimension. For a theory of brain function it would be insufficient to consider only one aspect of information processing, while disregarding all other aspects as noise. For a complete model of the functioning of the brain it seems important to also take into account the noise which can be defined as such only with regard to a specific processing of information, and thus address the interconnectedness of different instances of information processing. However, even in such a case it is difficult to find something like a general code. It seems therefore appropriate to emphasize precisely this fact of the non-existence of a general code of the nervous system as its most characteristic feature and to conceptualize this phenomenon. For this purpose Sigmund Freud's comments on the subject of free energy seem to be very useful.

We can thus distinguish between the basic provision of energy for the nervous system which can be fine-tuned by the circulatory system and the energetic coupling of information (it is due to this energetic coupling that the nervous system requires the provision of energy). It would be insufficient to not look for any codes in the pulse sequences of the nervous systems, as is suggested by some metaphors of constructivism, e.g. when Maturana points out that the effect of a virus depends on the characteristics of the host cell, as though the characteristics of the virus itself were of no importance. Such constructivist metaphors might lead us to neglect the fact that the type of external stimulation of the organism is by no means irrelevant. It is precisely the interplay of attempted coding and attempted self-stabilization within a system that can perhaps be illustrated best by using the concept of free energy.

If we start from the fact that the actual moment at which a nerve pulse sequence occurs is important for the encoding, then any variation in the starting point and temporal sequence, for whatever reason it may occur, will result in a failure to produce a constellation of information. In view of this fact we may introduce the assumption that the signals of the failed information constellation are available for further processing as "free energy". It is not the case that simply falsified information will be drawn from this situation, but instead the free energy in the nervous system is available for new constructions with the signals at hand, whereby the energy is never so freely uncoupled from the information that the "free energy" would be randomly available. This concept has the advantage that changes in the timing within the nervous system, which are viewed as problematic by coherence theory, are not simply deficits in information processing, but are in fact what makes the freedom, flexibility and constructiveness of the nervous system possible. In this context it should be pointed out once again that this constructiveness does not indicate randomness, although in extreme instances this may also be available to the organism, albeit with considerable consequences for its structure, state and behavior.

In contrast to the constructivist view that addresses only the self-organization in opposition to the concepts of information and program (see Maturana), this model offers a greater range of possibilities:

1. The energetic stabilization of information processing can be described as a process in which energy (within the general additional supply of energy) can be decoupled from the information processes themselves.

1.1 The introduction of the concept of free energy makes it possible to link neuroscientific models with considerations of the dynamics of instincts and psychoanalysis.

1.2 The concept of free energy allows us to ascribe states that cannot be interpreted as elaborated code not simply to chaos but to view the energetic dimension as constitutive for the course of the cognitive and emotional processes.

2. Phenomena of synchronization in nervous systems can now be viewed as attempts to introduce temporal parameters into the signal processing so that free energy can once again appear to be coupled to information processing. In this sense the concept of discharge, which is of considerable importance in psychoanalysis, might be reconstructed as a process of synchronization. Precisely because the nervous system does not have a clock, it searches for ways of rhythmization which can occur either through external rythms (techno music is an extreme example) or through mental concentration, but only in exceptional cases through aggression expressed by movement. In terms of the theory of discharge this means that it has at its disposal many possibilities other than that of motor manifestations. What can be the significance of introducing the category of free energy into the considerations of the continuity of neuroscience and information theory?

1. For those searching for a code of the nervous system it will become clear that the misleading assumption of a constant clock and time quanta can be replaced by a better concept in which the temporal structure is not the frame of reference for the information processes, but on the contrary, the characteristics of information processing contribute to the equalization of the temporal parameters and rhythms (synchronizations) of the nervous system due to its energetic coupling. This model therefore permits a clarification of that which is often wrongly presupposed by other models.

2. Of course, precisely placed information in the nervous system can also be traumatic if it ends up in a structural part of the system which shows an increase in energy. The reason that this information ends up in these regions of the brain (e.g., the coupling of an information in the area of the hippocampus with an activation of the septum) can be explained to some extent by the concept of free energy as a decoupling of information and energy. In this way several pathological phenomena become evident: the accumulation of free energy makes clear why the information processing is concentrated precisely in these areas of the brain (the trauma is an excess of energy over information). At this point the possibility of freedom also becomes clear. It is not an absolute freedom. However, if it is aware of the fact that it depends on the energetic decouplings and on the other hand, must for this very reason, also be synchronization, it can develop better and expand its range of decisions than if it were to view its neuronal basis merely as determinate, hardly taking into account the specific properties of the external stimuli.

REFERENCES

Arbib, M. (Ed.) (1995). *The Handbook of Brain Theory and Neural Networks*, Cambridge, Mass.: MIT Press.

Freeman, W. J., Skarda, C. A.: (1985). Spatial EEG Patterns, Nonlinear Dynamics and Perception: the Neo-Sherringtonian View. *Brain Research Reviews*, 10, 147–175.

Freud, S. (1966). Project for a Scientific Psychology. In: *Standard Edition*, vol. I, London: Hogarth Press, pp. 177–387.

Kandel, E. R. et al. (Eds.) (1995). *Essentials of Neural Science and Behavior*, Stamford, Conn: Appleton & Lange.

Maturana, H. R. (1980). *Autopoiesis and Cognition: The Realization of the Living*, Dordrecht, Holland; Boston: D. Reidel Pub. Co.

Shannon, C. E., Weaver, W. (1964). *The Mathematical Theory of Communication*, Urbana: University of Illinois Press.

List of Facsimiles

Between pages 10 and 11
Freud, S. (1877). Über das Syrskische Organ etc. In: *Sitzungsberichte der kaiserlichen Akademie der Wissenschaften. Mathematisch-naturwissenschaftliche Classe*, vol. LXXV, No. 1, between pp. 430 and 431.

Between pages 22 and 23
Freud, S. (1879). Über Spinalganglien und Rückenmark des Petromyzon; Tafel 1. In: *Sitzungsberichte der kaiserlichen Akademie der Wissenschaften. Mathematisch-naturwissenschaftliche Classe*, vol. LXXVIII, Nos. 1–5, between pp. 160 and 161.

Between pages 36 and 37
Freud, S. (1879). Über Spinalganglien und Rückenmark des Petromyzon; Tafel 2. In: *Sitzungsberichte der kaiserlichen Akademie der Wissenschaften. Mathematisch-naturwissenschaftliche Classe*, vol. LXXVIII, Nos. 1–5, between pp. 160 and 161.

Between pages 46 and 47
Freud, S. (1879). Über Spinalganglien und Rückenmark des Petromyzon; Tafel 3. In: *Sitzungsberichte der kaiserlichen Akademie der Wissenschaften. Mathematisch-naturwissenschaftliche Classe*, vol. LXXVIII, Nos. 1–5, between pp. 160 and 161.

Between pages 56 and 57
Freud, S. (1879). Über Spinalganglien und Rückenmark des Petromyzon; Tafel 4. In: *Sitzungsberichte der kaiserlichen Akademie der Wissenschaften. Mathematisch-naturwissenschaftliche Classe*, vol. LXXVIII, Nos. 1–5, between pp. 160 and 161.

Between pages 86 and 87
Freud, S. (1882). Über den Bau der Nervenfasern und Nervenzellen beim Flusskrebs. In: *Sitzungsberichte der kaiserlichen Akademie der Wissenschaften. Mathematisch-naturwissenschaftliche Classe*, vol. LXXV, Nos. 1–2, between pp. 46 and 47.

Between pages 102 and 103
Freud, S. (1877). Ursprung der hinteren Nervenwurzeln etc. In: *Sitzungsberichte der kaiserlichen Akademie der Wissenschaften. Mathematisch-naturwissenschaftliche Classe*, vol. LXXV, Nos. 1–5, between pp. 26 and 27.

Copyright for illustrations by S. Freud: A. W. Freud *et al.*/Mark Paterson & Associates

Index of Proper Names

Abraham, Karl 11
Adams, K. S. 90, 99
Aldridge, Victor J. 35
Allen, S. 100
Amacher, Peter 60, 71, 83
Antoni, M. H. 99
Arbib, M. 108
Aspendorf, Jens W. 89, 99
Averill, J. R. 100

Babinski, Joseph 17
Bacher, Richard 28, 34
Bakhtin, Michael M. 47, 54, 55
Barolin, Gerhard 31, 34
Bartlett, Frederic 20
Bauer, Herbert 29, 34, 35
Beall, S.K. 96, 100
Beck, Adolf T. 28
Bell, A.J. 89, 99
Benjamin, Walter 46
Berger, Hans 26, 34
Berman, K. F. 101
Bergmann, Gustav von 40
Bernays, Martha 43
Bernfeld, Siegfried 24, 28, 34, 44, 60, 62
Bernheim, Hippolyte 16
Billroth, Christian Albert 43
Binet, Alfred 87
Bonaparte, Marie 42
Borck, Cornelius 8
Breuer, Josef 24, 25, 34, 39, 43, 97, 99
Broca, Paul 12
Brown, William A. 90, 99
Brücke, Ernst Wilhelm von 8, 11, 23, 24, 38, 43, 44, 45, 59, 60, 61, 72, 79
Brun, Rudolf 40, 62
Burian, Kurt 27
Burke, Kenneth 59

Cajal, Ramón y Santiago 12, 47
Cascione, R. 100
Cassirer, Ernst 49, 50
Cassirer-Bernfeld, Suzanne 24, 34, 44, 60
Caton, Richard 28
Cerro, Manuel de 62
Charcot, Jean Martin 13, 16, 17, 45, 87
Claus, Carl 38
Cohen, F. 101
Cohen, J. 34
Colder, M. 96, 100
Cole, Michael 21
Cole, Sheila 21
Cook, H. 89, 99
Cooper, Richard 35
Crowne, Douglas P. 89, 99
Cybulski 28

Damasio, A. R. 55
Darkshevich, Liweri Ossipovich 41
Darwin, Charles Robert 11, 12, 13, 21
Daston, Loraine 62
Davidson, R.J. 88, 89, 90, 92, 93, 99, 101
Delaney, R. 100
Derrida, Jacques 85
Dimond, Stuart J. 92, 93, 99
Dohrer, Maria 83
Doyle, B. 101
DuBois-Reymond, Emil 23, 44

Eagle, Morris N. 8, 98, 99
Economo, Constantin von 44
Edelman, Gerald M. 20, 21
Ekman, P. 99
Ellenberger, Henry F. 38
Emde, Robert N. 54
Emmons, R. A. 100
Engebretson, T. O. 99

Esterling, B. A. 90, 94, 96, 99
Exner, Sigmund 15, 43, 60, 71

Fallend, Karl 28, 34
Fancher, Raymond E. 71
Farrington, L. 92, 93, 99
Fechner, Gustav Theodor 26, 34
Feitelberg, Sergei 28
Feuchtersleben, Ernst von 43
Flechsig, Paul 12
Fleischl-Marxow, Ernst von 43, 60
Fliess, Wilhelm 7, 17, 18, 19, 40, 42, 46, 47
Forrester, John 57, 68
Fox, N.A. 93, 99
Freeman, Walter J. 55, 104, 108
Freud, Anna 42
Freud, Ernst 60
Freud, Ernst L. 43
Freud, Lucie 60
Fuchs, S. H. 51, 55

Galin, D. 92, 100
Galison, Peter 62
Geppert, Sebastian 68
Gerstmann, J. 52
Gestring, Gideon 34
Gestring, Gidon 27, 31
Gill, Merton M. 19, 22, 98, 100
Gittler, Georg 35
Glaser, R. 97, 101
Goethe, Johann Wolfgang von 11
Goldberg, S. 99
Goldstein, Kurt 8, 47, 48, 49, 50, 51, 52, 55, 92, 100
Gradman, A. 100
Grashey, Hubert G. 66, 67
Gregory, Richard 20
Gross, J. J. 91, 100
Grubrich-Simitis, Ilse 24, 43, 44, 60, 84
Gur, E. C. 93, 100
Gur, R. E. 93, 100
Guttmann, Giselher 9, 27, 29, 34

Hartmann, Heinz 51
Head, Henry 48, 49
Hebb, Donald 19
Hegeson, V.S. 101
Helmholtz, Hermann von 23, 59

Heynick, Frank 71
Holmes, Gordon 21
Holquist, M. 55
Holt, Robert R. 98, 100, 101
Holzman, P. S. 100
Huang, S. S. 100
Hughes, C.F. 97, 100
Humboldt, Wilhelm von 49
Husserl, Edmund 50

Irle, M. 34
Izard, C.E. 99

Jackson, Hughlings John 8, 13, 14, 15, 16, 21, 50
Jamner, L. D. 90, 100
Janet, Pierre 87
Jelliffe, Smith Ely 21
Jensen, M. R. 90, 100
Johnson, P. 92, 93, 99
Jones, Ernest 21, 60

Kagan, J. 99
Kandel, Eric R. 108
Kant, Immanuel 50
Karrer, R. 34
Kassowitz, Max 38, 39, 40, 41, 42, 45
Kästle, Oswald U. 70
Keller, A.E.S. 90, 99
Kerns, R 100
Kerr, J. 99
Kiecolt-Glaser, J. K 97, 101
King, L. A. 90, 100
Kohut, Heinz 54
Koriat, A. 91, 100
Korunka, Christian 34
Krafft-Ebing, Richard von 44
Kris, Ernst 42
Kumar, M. 99
Kundrat, Hans 37

Lakatos, Imra 98, 100
Lane, Robert D. 89, 100
Lang, Johannes 40
Langer, G. 34
Laudan, L. 99
Lazarus, R. S. 100
Leigh, H. 90, 100

Index of Proper Names

Leodolter, Michael 34, 35
Leodolter, Ulrich 35
Lesky, Erna 45
Leupold-Löwenthal, Harald 8, 45
Levenson, R. W. 91, 100
Leventhal, H. 90, 101
Levin, F. 52
Levine, J. 90, 95, 100
Lichtheim 68
Linke, Detlef B. 9
Love, R. 90, 101
Ludwig, Carl 23
Luria, Alexander R. 15, 21, 22

Mahony, Patrick J. 59
Marlowe, D. H. 89, 99
Maturana, Humberto R. 108
McCallum, William C. 35
Melman, R. 100
Meynert, Theodor 12, 14, 38, 43, 44, 45, 46, 60, 79, 83
Mitchell, Weir 17
Modell, Arnold 19, 20, 22, 55
Monakow, Constantin von 40
Muir, R. 99
Müller, Johannes 23, 44
Musgrave, A. 100

Niaura, R. 99
Nietzsche, Friedrich 20
Nothnagel, Hermann 40, 41, 43, 44, 45

Obersteiner, Heinrich 44
O' Heeron, R. C. 97, 100
Oppenheim, Hermann 41
Ottenthal, Rudolf von 26

Palesky, J. 101
Patterson, Gordon 59
Pennebaker, James W. 96, 97, 100, 101
Pines, Malcolm 8
Pribram, Karl 18, 19, 22, 28, 34

R., Elisabeth von 42
Ramachandran, V. S. 20
Rebert, Ch. 34
Reed, G. M. 101
Reicheneder, Johann G. 70

Reichmayr, Johannes 28, 34
Reinhard 39
Reissner 24
Rickman, John 55
Rie, Oskar 39, 40, 41
Riese, Walter 49, 50, 55
Rohracher, Hubert 26, 27, 34, 35
Rokitansky, Carl 43, 44
Rollet, Alexander 43
Rolter, W. L. 55
Rosenthal, Moriz 43
Rosenzweig, H. 34
Rubinstein, Benjamin 98, 101
Rusinov, Vladimir 28

Sacks, Oliver W. 8, 55
Saling, Michael 21, 22, 71
Saussure, Ferdinand de 47, 54
Scherer, K. 89, 99
Schilder, Paul 8, 47, 48, 49, 50, 51, 52, 55
Schneiderman, N. 99
Scholz, Franz 37, 38
Schönau, Walter 59
Schore, Allan N. 94, 101
Schott, Heinz G. 58
Schwartz, George E. 88, 90, 95, 100, 101
Shannon, C. E. 108
Sharp, L. K. 96, 100
Shaskan, D. A. 55
Shaw, Robert E. 95, 101
Sherrington, Charles 19
Shotter, John 54, 55
Signorelli, Luca 73
Simmel, Marianne 50
Sinz, R. 34
Sirota, A.D. 99
Skarda, C.A. 108
Skoda, Joseph 43
Skokan, L.A. 101
Smith, Roger 48, 55
Smith, S. 100
Solms, Mark 21, 22, 71
Sontana, A. 100
Spencer, Herbert 13
Staehlin, Peter 40
Stengel, Ervine 49, 68
Stern, Daniel 52
Stewart, Walter 48

Strachey, James 46
Stricker, Salomon 38, 60
Strümpell, Adolf 39
Sulloway, Frank J. 22, 71

Taylor, James 21
Taylor, S. E. 89, 101
Tomarken, A. J. 89, 90, 101
Triarhou, Lazaros C. 62
Tueting, P. 34
Türckheim 43
Türk, Ludwig 43

Vigotsky, Lev S. 51, 54
Villaret, Otto 12, 42, 70
Vitouch, Oliver 30, 35
Volosinov 54

Wagner-Jauregg, Julius R. von 44
Waldeyer, Anton 12
Walter, Grey 28, 35
Ward, S. E. 90, 101
Warrenburg, S. 100
Weaver, W. 108
Weinberger, Daniel R. 88, 89, 94, 95, 96, 101
Weiss, Nathan 43
Wernicke, Carl 47, 48, 51, 65, 66
West, M. 90, 99
Wilbrand, Franz Julius 39
Winter, Allan L. 35
Wlashe, F. M.R. 21

Zajonc, R. E. 99
Zec, R. F. 101
Ziferstein 51

Biographies

At the symposion: Detlef B. Linke, Giselher Guttmann, Werner Welzig, Cornelius Borck, Inge Scholz-Strasser, August Ruhs, Harald Leupold-Löwenthal, Malcolm Pines, Morris N. Eagle, Oliver W. Sacks (from left to right)

Photograph by Karl Reiberger

CORNELIUS BORCK (Berlin) studied medicine, philosophy, history of medicine and theology at the universities of Hamburg, Heidelberg and Berlin. After finishing a PhD in neurosciences at *Imperial College, University of London*, he started a historical study on the construction of electroencephalography at the Institute for Science and Technology Studies of the *University of Bielefeld*. He is currently working at the *Max-Planck-Institute for the History of Science* in *Berlin*. Cornelius Borck has published in the field of epilepsy research, on the history of electrophysiological brain research and on visualization strategies.

MORRIS N. EAGLE (New York) received his Ph.D. in clinical psychology from *New York University*. He was head of the *Program in Clinical Psychology* at *York University* and has taught at Cambridge, Berkeley and Pittsburgh. From 1995 to 1996 he was President of the Division of Psychoanalysis of the American Psychological Association. Since 1995 he has been a professor at the *Derner Institute of Advanced Psychological Studies of Adelphi University, New York*. In addition to his teaching and research, Morris Eagle works as a psychoanalyst in private practice. He is the author of numerous publications on psychoanalysis.

GISELHER GUTTMANN (Vienna) studied psychology, zoology and philosophy at the University of Vienna. Since 1968 he has been professor of *Psychology* at the *Department of Psychology* of the *University of Vienna*. In 1975 he was appointed Dean of the *Philosophical Faculty* and in 1976 Dean of the *"Grund- und Integrativwissenschaftliche Fakultät"*. Since 1983 G. Guttmann has been a fellow of the Austrian Academy of Sciences and from 1994 on has also served as Distinguished Professor at the *International Academy of Philosophy in the Principality of Liechtenstein*. He has published extensively on general and experimental psychology and neuropsychology.

HARALD LEUPOLD-LÖWENTHAL (Vienna) is Dozent for Psychotherapy and Psychoanalysis at the *University of Vienna* and worked as a psychiatrist and neurologist. From 1974 to 1981 he was Chairman of the Vienna Psychoanalytic Society and is still active as a training analyst. Harald Leupold-Löwenthal has served as Secretary of the *International Psychoanalytic Association* and is currently Editorial Advisor of *Sigmund Freud Copyrights* and President of the *Sigmund Freud Society*. He lives in Vienna as a psychoanalyst in private practice. His numerous publications address topics of cultural history and psychoanalysis.

DETLEF B. LINKE (Bonn) studied medicine, philosophy, communications science and phonetics at the University of Bonn. In 1982 he was appointed Professor of *Clinical Neurophysiology and Neurosurgical Rehabilitation* at the *University of Bonn*. He was co-founder of the Center for Gerontology at the same university and also of the journal *Ethica. Wissenschaft und Verantwortung*. He is currently Vice President of the *Society for the Philosophical Study of Genocide and the Holocaust* of the *City University of New York* and is a member of the graduate college *Intercultural Religious Studies* of the *Deutsche Forschungsgemeinschaft* as well as a member of the *Academy of Ethics in Medicine*. He is a member of the *Neuroscientific Advisory Board of the New York Psychoanalytic Institute* and the author of numerous publications on epilepsy, neurology and neurophysiology.

MALCOLM PINES (London) has a private practice for psychoanalysis and group analysis. He is a member of the *Group-Analytic Practice, London*, and Chairman of the *Curriculum Committee, Institute of Group Analysis, London*. He previously worked as a psychotherapist and psychiatrist at *Cassel Hospital, St. George's Hospital, Maudsley Hospital* and at the *Tavistock Clinic*. He has published numerous works on psychoanalysis and group psychotherapy.

OLIVER W. SACKS (New York) studied physiology, biology and biochemistry at the universities of London and Oxford. He works as a neurologist and psychiatrist at several institutions: he is Professor of Neurology at the *Albert Einstein College of Medicine, New York*, and Scientific Advisor at the *Institute for Music and Neurologic Function, Beth Abraham Hospital, New York*. His publications have been translated into numerous languages and have won him several international awards, among them the *Hawthornden Prize* for *Awakenings* in 1975. He is a fellow of the American Academy of Arts and Letters. He has published extensively on neurology and neuropsychiatry and has written seven books including *The Man Who Mistook his Wife for a Hat* and *An Anthropologist on Mars*.

INGE SCHOLZ-STRASSER (Vienna) studied history and philosophy at the *University of Vienna*. Since 1987 she has been Executive Secretary of the *Sigmund Freud Society* and in this function also director of the Sigmund Freud Museum in Vienna. She has edited and contributed her own articles to several publications on Viennese history and psychoanalysis.